Basic Gas Chromatography

TECHNIQUES IN ANALYTICAL CHEMISTRY SERIES

Basic Gas Chromatography

HAROLD M. McNAIR, Ph.D.
Virginia Polytechnic Institute and State University

JAMES M. MILLER, Ph.D.
Drew University

A Wiley-Interscience Publication

JOHN WILEY & SONS, INC.

New York • Chichester • Weinheim • Brisbane • Singapore • Toronto

Cover: GC photograph reprinted by permission of J&W Scientific, Inc. Photograph appeared on the cover of *LC/GC* magazine, May 1997.

This text is printed on acid-free paper. ∞

Library of Congress Cataloging in Publication Data:

McNair, Harold Monroe, 1933–
        Basic gas chromatography / Harold M. McNair, James M. Miller.
            p.   cm.—(Techniques in analytical chemistry series)
        "A Wiley-Interscience publication."
        Includes bibliographical references and index.
        ISBN 0-471-17260-X (alk. paper).—ISBN 0-471-17261-8 (pbk. : alk. paper)
        1. Gas chromatography.   I. Miller, James M., 1933–
    II. Title.   III. Series.
    QD79.C45M425   1997
543'.0896—dc21   97-18151
                                                        CIP

Printed in the United States of America

10   9   8   7   6   5   4   3   2   1

Contents

Series Preface

Titles in the *Techniques in Analytical Chemistry Series* address current techniques in general use by analytical laboratories. The series intends to serve a broad audience of both practitioners and clients of chemical analysis. This audience includes not only analytical chemists but also professionals in other areas of chemistry and in other disciplines relying on information derived from chemical analysis. Thus, the series should be useful to both laboratory and management personnel.

Written for readers with varying levels of education and laboratory expertise, titles in the series do not presume prior knowledge of the subject, and guide the reader step-by-step through each technique. Numerous applications and diagrams emphasize a practical, applied approach to chemical analysis.

The specific objectives of the series are:

- to provide the reader with overviews of methods of analysis that include a basic introduction to principles but emphasize such practical issues as technique selection, sample preparation, measurement procedures, data analysis, quality control and quality assurance;
- to give the reader a sense of the capabilities and limitations of each technique, and a feel for its applicability to specific problems;
- to cover the wide range of useful techniques, from mature ones to newer methods that are coming into common use; and
- to communicate practical information in a readable, comprehensible style. Readers from the technician through the PhD scientist or labora-

tory manager should come away with ease and confidence about the use of the techniques.

Forthcoming books in the *Techniques in Analytical Chemistry Series* will cover a variety of techniques including chemometric methods, biosensors, surface and interface analysis, measurements in biological systems, inductively coupled plasma-mass spectrometry, gas chromatography–mass spectrometry, Fourier transform infrared spectroscopy, and other significant topics. The editors welcome your comments and suggestions regarding current and future titles, and hope you find the series useful.

FRANK A. SETTLE

Lexington, VA

Preface

A series of books on the *Techniques in Analytical Chemistry* would be incomplete without a volume on gas chromatography (GC), undoubtedly the most widely used analytical technique. Over 40 years in development, GC has become a mature method of analysis and one that is not likely to fade in popularity.

In the early years of development of GC, many books were written to inform analysts of the latest developments. Few of them have been kept up-to-date and few new ones have appeared, so that a satisfactory single introductory text does not exist. This book attempts to meet that need. It is based in part on the earlier work by the same title, *Basic Gas Chromatography,* co-authored by McNair and Bonelli and published by Varian Instruments. Some material is also drawn from the earlier Wiley book by Miller, *Chromatography: Concepts and Contrasts.*

We have attempted to write a brief, basic, introduction to GC following the objectives for titles in this series. It should appeal to readers with varying levels of education and emphasizes a practical, applied approach to the subject. Some background in chemistry is required: mainly general organic chemistry and some physical chemistry. For use in formal class work, the book should be suitable for undergraduate analytical chemistry courses and for intensive short courses of the type offered by the American Chemical Society and others. Analysts entering the field should find it indispensable, and industrial chemists working in GC should find it a useful reference and guide.

Because the IUPAC has recently published its nomenclature recommendations for chromatography, we have tried to use them consistently to promote a unified set of definitions and symbols. Also, we have endeavored to write in such a way that the book would have the characteristics of a single author, a style especially important for beginners in the field. Otherwise, the content and coverage are appropriately conventional.

While open tubular (OT) columns are the most popular type, both open tubular and packed columns are treated throughout, and their advantages, disadvantages, and applications are contrasted. In addition, special chapters are devoted to each type of column. Chapter 2 introduces the basic instrumentation and Chapter 7 elaborates on detectors. Other chapters cover stationary phases (Chapter 4), qualitative and quantitative analysis (Chapter 8), programmed temperature (Chapter 9), and troubleshooting (Chapter 11). Chapter 10 briefly covers the important special topics of GC–MS, derivatization, chiral analysis, headspace sampling, and solid phase microextraction (SPME) for GC analysis.

We would like to express our appreciation to our former professors and many colleagues who have in one way or another aided and encouraged us and to those students who, over the years, have provided critical comments that have challenged us to improve both our knowledge and communication skills.

HAROLD M. MCNAIR
JAMES M. MILLER

Basic Gas Chromatography

1. Introduction

It is hard to imagine an organic analytical laboratory without a gas chromatograph. In a very short time gas chromatography, GC, has become the premier technique for separation and analysis of volatile compounds. It has been used to analyze gases, liquids, and solids—the latter usually dissolved in volatile solvents. Both organic and inorganic materials can be analyzed, and molecular weights can range from 2 to over 1,000 Daltons.

Gas chromatographs are the most widely used analytical instruments in the world [1]. Efficient capillary columns provide high resolution, separating more than 450 components in coffee aroma, for example, or the components in a complex natural product like peppermint oil (see Fig. 1.1). Sensitive detectors like the flame-ionization detector can quantitate 50 ppb of organic compounds with a relative standard deviation of about 5%. Automated systems can handle more than 100 samples per day with minimum down time, and all of this can be accomplished with an investment of less than $20,000.

A BRIEF HISTORY

Chromatography began at the turn of the century when Ramsey [2] separated mixtures of gases and vapors on adsorbents like charcoal and Michael Tswett [3] separated plant pigments by liquid chromatography. Tswett is credited as being the "father of chromatography" principally because he

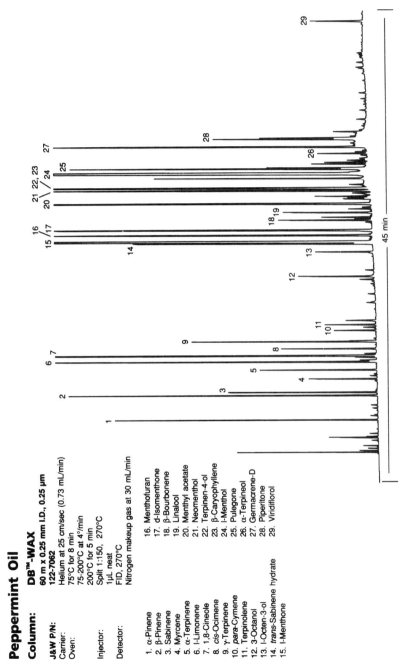

Peppermint Oil

Column: DB™-WAX
60 m x 0.25 mm I.D., 0.25 μm

J&W P/N: 122-7062
Carrier: Helium at 25 cm/sec (0.73 mL/min)
Oven: 75°C for 8 min
75-200°C at 4°/min
200°C for 5 min
Injector: 1μL neat
Split 1:150, 270°C
Detector: FID, 270°C
Nitrogen makeup gas at 30 mL/min

1. α-Pinene
2. β-Pinene
3. Sabinene
4. Myrcene
5. α-Terpinene
6. l-Limonene
7. 1,8-Cineole
8. cis-Ocimene
9. γ-Terpinene
10. para-Cymene
11. Terpinolene
12. 3-Octanol
13. l-Octen-3-ol
14. trans-Sabinene hydrate
15. l-Menthone

16. Menthofuran
17. d-Isomenthone
18. β-Bourbonene
19. Linalool
20. Menthyl acetate
21. Neomenthol
22. Terpinen-4-ol
23. β-Caryophyllene
24. l-Menthol
25. Pulegone
26. α-Terpineol
27. Germacrene-D
28. Piperitone
29. Viridiflorol

Fig. 1.1. Typical gas chromatographic separation showing the high efficiency of this method. Courtesy of J & W Scientific, Inc.

2

coined the term chromatography (literally color writing) and scientifically described the process. His paper has been translated into English and republished [4] because of its importance to the field. Today, of course, most chromatographic analyses are performed on materials that are not colored.

Gas chromatography is that form of chromatography in which a gas is the moving phase. The important seminal work was first published in 1952 [5] when Martin and his co-worker James acted on a suggestion made 11 years earlier by Martin himself in a Nobel-prize winning paper on partition chromatography [6]. It was quickly discovered that GC was simple, fast, and applicable to the separation of many volatile materials—especially petrochemicals, for which distillation was the preferred method of separation at that time. Theories describing the process were readily tested and led to still more advanced theories. Simultaneously the demand for instruments gave rise to a new industry that responded quickly by developing new gas chromatographs with improved capabilities.

The development of chromatography in all of its forms has been thoroughly explored by Ettre who has authored nearly 50 publications on chromatographic history. Three of the most relevant articles are: one focused on the work of Tswett, Martin, Synge, and James [7]; one emphasizing the development of GC instruments [8]; and the third, which contained over 200 references on the overall development of chromatography [9].

Today GC is a mature technique and a very important one. The worldwide market for GC instruments is estimated to be about $1 billion or over 30,000 instruments annually.

DEFINITIONS

In order to define chromatography adequately, a few terms and symbols need to be introduced, but the next chapter is the *main* source of information on definitions and symbols.

Chromatography

Chromatography is a separation method in which the components of a sample partition between two phases: one of these phases is a stationary bed with a large surface area, and the other is a gas which percolates through the stationary bed. The sample is vaporized and carried by the mobile gas phase (the *carrier gas*) through the column. Samples partition (equilibrate) into the stationary liquid phase, based on their solubilities at the given temperature. The components of the sample (called solutes or analytes) separate from one another based on their *relative* vapor pressures and affinities for the stationary bed. This type of chromatographic process is called *elution.*

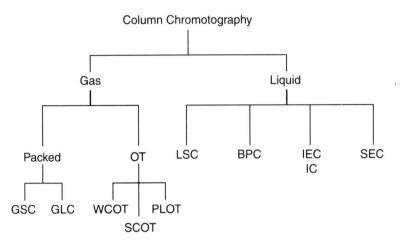

Fig. 1.2. Classification of chromatographic methods.

The "official" definitions of the International Union of Pure and Applied Chemistry (IUPAC) are: "Chromatography is a physical method of separation in which the components to be separated are distributed between two phases, one of which is stationary (stationary phase) while the other (the mobile phase) moves in a definite direction. Elution chromatography is a procedure in which the mobile phase is continuously passed through or along the chromatographic bed and the sample is fed into the system as a finite slug" [10].

The various chromatographic processes are named according to the physical state of the mobile phase. Thus, in gas chromatography (GC), the mobile phase is a *gas,* and in liquid chromatography (LC) the mobile phase is a *liquid.* A subclassification is made according to the state of the stationary phase. If the stationary phase is a solid, the GC technique is called gas-solid chromatography (GSC), and if it is a liquid, gas-liquid chromatography (GLC).

Obviously, the use of a gas for the mobile phase requires that the system be contained and leak-free, and this is accomplished with a glass or metal tube referred to as the column. Since the column contains the stationary phase, it is common to name the column by specifying the stationary phase, and to use these two terms interchangeably. For example, one can speak about an OV-101[a] column, which means that the stationary liquid phase is OV-101 (see Chapter 4).

A complete classification scheme is shown in Figure 1.2. Note especially the names used to describe the open tubular (OT) GC columns and the LC columns; they do not conform to the guidelines just presented. However,

[a] OV designates the trademarked stationary liquid phases of the Ohio Valley Specialty Chemical Company of Marietta, Ohio.

all forms of GC can be included in two subdivisions, GLC and GSC; some of the capillary columns represent GLC and others, GSC. Of the two major types, GLC is by far the more widely used, and consequently, it receives the greater attention in this work.

The Chromatographic Process

Figure 1.3 is a schematic representation of the chromatographic process. The horizontal lines represent the column; each line is like a snapshot of the process at a different time (increasing in time from top to bottom). In the first snapshot, the sample, composed of components A and B, is introduced onto the column in a narrow zone. It is then carried through the column (from left to right) by the mobile phase.

Each component partitions between the two phases, as shown by the distributions or peaks above and below the line. Peaks above the line represent the amount of a particular component in the mobile phase, and peaks below the line the amount in the stationary phase. Component A

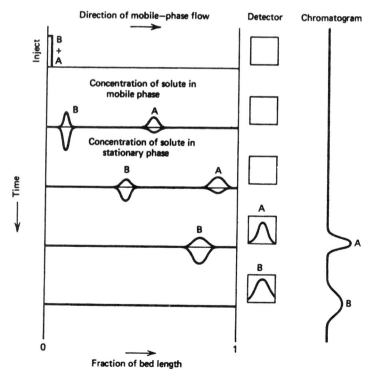

Fig. 1.3. Schematic representation of the chromatographic process. From Miller, J. M., *Chromatography: Concepts and Contrasts,* John Wiley & Sons, Inc., New York, 1987, p. 7. Reproduced courtesy of John Wiley & Sons, Inc.

has a greater distribution in the mobile phase and as a consequence it is carried down the column faster than component B, which spends more of its time in the stationary phase. Thus, separation of A from B occurs as they travel through the column. Eventually the components leave the column and pass through the detector as shown. The output signal of the detector gives rise to a *chromatogram* shown at the right side of Figure 1.3.

Note that the figure shows how an individual chromatographic peak widens or broadens as it goes through the chromatographic process. The exact extent of this broadening, which results from the kinetic processes at work during chromatography, will be discussed in Chapter 3.

The tendency of a given component to be attracted to the stationary phase is expressed in chemical terms as an equilibrium constant called the *distribution constant,* K_c, sometimes also called the partition coefficient. The distribution constant is similar in principle to the partition coefficient that controls a liquid-liquid extraction. In chromatography, the greater the value of the constant, the greater the attraction to the stationary phase.

Alternatively, the attraction can be classified relative to the *type* of *sorption* by the solute. Sorption on the surface of the stationary phase is called *ad*sorption and sorption into the bulk of a stationary liquid phase is called *ab*sorption. These terms are depicted in comical fashion in Figure 1.4. However, most chromatographers use the term *partition* to describe the absorption process. Thus they speak about adsorption on the surface of the stationary phase and partitioning as passing into the bulk of the stationary phase. Usually one of these processes is dominant for a given column, but both can be present.

The distribution constant provides a numerical value for the total sorption by a solute *on* or *in* the stationary phase. As such, it expresses the extent of interaction and regulates the movement of solutes through the chromatographic bed. In summary, differences in distribution constants

ABsorption   **ADsorption**

Fig. 1.4. Comical illustration of the difference between absorption (partition) and adsorption. From Miller, J. M., *Chromatography: Concepts and Contrasts,* John Wiley & Sons, Inc., New York, 1987, p. 8. Reproduced courtesy of John Wiley & Sons, Inc.

(parameters controlled by thermodynamics) effect a chromatographic separation.

Some Chromatographic Terms and Symbols

The IUPAC has attempted to codify chromatographic terms, symbols, and definitions for all forms of chromatography [10], and their recommendations will be used in this book. However, until the IUPAC publication in 1993, uniformity did not exist and some confusion may result from reading older publications. Table 1.1 compares some older conventions with the new IUPAC recommendations.

The distribution constant, K_c, has just been discussed as the controlling factor in the partitioning equilibrium between a solute and the stationary phase. It is defined as the concentration of the solute A in the stationary phase divided by its concentration in the mobile phase.

$$K_c = \frac{[A]_S}{[A]_M} \tag{1}$$

This constant is a true thermodynamic value which is temperature dependent; it expresses the relative tendency of a solute to distribute itself between the two phases. Differences in distribution constants result in differential migration rates of solutes through a column.

Figure 1.5 shows a typical chromatogram for a single solute, A, with an additional small peak early in the chromatogram. Solutes like A are retained by the column and are characterized by their *retention volumes*, V_R; the retention volume for solute A is depicted in the figure as the distance from the point of injection to the peak maximum. It is the volume of carrier gas

TABLE 1.1 Chromatographic Terms and Symbols

Symbol and Name Recommended by the IUPAC*	Other Symbols and Names in Use
K_c Distribution constant (for GLC)	K_p Partition coefficient
	K_D Distribution coefficient
k Retention factor	k' Capacity factor; capacity ratio; partition ratio
N Plate number	n Theoretical plate number; no. of theoretical plates
H Plate height	HETP Height equivalent to one theoretical plate
R Retardation factor (in columns)	R_R Retention ratio
R_s Peak resolution	R
α Separation factor	Selectivity; solvent efficiency
t_R Retention time	
V_R Retention volume	
V_M Hold-up volume	Volume of the mobile phase; V_G volume of the gas phase; V_O void volume; dead volume

*Source: Data taken from Ref. 10.

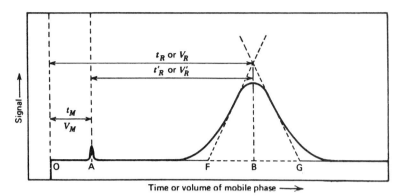

Fig. 1.5. Typical chromatogram. From Miller, J. M., *Chromatography: Concepts and Contrasts,* John Wiley & Sons, Inc., New York, 1987, p. 8. Reproduced courtesy of John Wiley & Sons, Inc.

necessary to elute solute A. This characteristic of a solute could also be specified by the retention time, t_R, if the column flow rate, F_c, were constant[b].

$$V_R = t_R \times F_c \qquad (2)$$

Unless specified otherwise, a constant flow rate is assumed and retention time is proportional to retention volume and both can be used to represent the same concept.

The small early peak represents a solute that does not sorb in the stationary phase—it passes straight through the column without stopping. In GC, this behavior is often shown by air or methane, and the peak is often called an air peak. The symbol V_M, sometimes called the hold-up volume or void volume, serves to measure the interstitial or interparticle volume of the column. Other IUPAC approved symbols include V_o and V_G, representing the volume of the mobile gas phase in the column. The term dead volume, while not recommended, is also widely used.

Equation 3, one of the fundamental chromatographic equations[c], relates the chromatographic *retention volume* to the theoretical distribution constant.

$$V_R = V_M + K_C V_S \qquad (3)$$

[b] Because the chromatographic column is under pressure, the carrier gas volume is small at the high-pressure inlet, but expands during passage through the column as the pressure decreases. This topic is discussed in Chapter 2.

[c] For a derivation of this equation, see: B. L. Karger, L. R. Snyder, and C. Horvath, *An Introduction to Separation Science,* Wiley, NY, 1973, pp. 131 and 166.

V represents a volume and the subscripts R, M, and S stand for retention, mobile, and stationary, respectively. V_M and V_S represent the volumes of mobile phase and stationary phase in the column respectively. The retention volume, V_R can be described by reference to Figure 1.5.

An understanding of the chromatographic process can be deduced by reexamining equation 3. The total volume of carrier gas that flows during the elution of a solute can be seen to be composed of two parts: the gas that fills the column or, alternatively, the volume through which the solute must pass in its journey through the column as represented by V_M, and, second, the volume of gas that flows while the solute is not moving but is stationary on or in the column bed. The latter is determined by the distribution constant (the solute's tendency to sorb) and the amount of stationary phase in the column, V_S. There are only two things a solute can do: move with the flow of mobile phase when it is in the mobile phase, or sorb into the stationary phase and remain immobile. The sum of these two effects is the total retention volume, V_R.

OVERVIEW: ADVANTAGES AND DISADVANTAGES

GC has several important advantages as summarized in the list below.

Advantages of Gas Chromatography

- Fast analysis, typically minutes
- Efficient, providing high resolution
- Sensitive, easily detecting ppm and often ppb
- Nondestructive, making possible on-line coupling; e.g., to mass spectrometer
- Highly accurate quantitative analysis, typical RSDs of 1–5%
- Requires small samples, typically μL
- Reliable and relatively simple
- Inexpensive

Chromatographers have always been interested in fast analyses, and GC has been the fastest of them all, with current commercial instrumentation permitting analyses in seconds. Figure 1.6 shows a traditional orange oil separation taking 40 minutes, a typical analysis time, and a comparable one completed in only 80 seconds using instrumentation specially designed for fast analysis.

The high efficiency of GC was evident in Figure 1.1. Efficiency can be expressed in plate numbers, and capillary columns typically have plate numbers of several hundred thousand. As one might expect, an informal competition seems to exist to see who can make the column with the

ORANGE OIL, 1000 PPM in ISOOCTANE

(a) **Industry Standard**
Conditions not reported

Components:
1 - Ethyl Butyrate
2 - Isomyl acetate
3 - (alpha)-pinene
4 - Myrcene
5 - Octanal
6 - *p*-Cymene
7 - Limonene
8 - Linalool
9 - 4-terpineol
10 - (alpha)-terpineol
11 - Decanal
12 - Neral
13 - Carvone
14 - Geranail
15 - Perillaldehyde
16 - Undecanal
17 - Dodecanal
18 - (alpha)-ionone
19 - Cadinene
20 - Int.

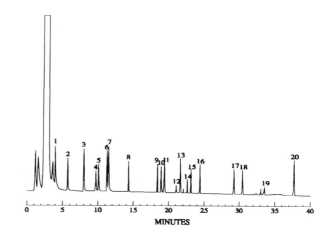

(b) **Flash-2D-GC**
Column: DB-5, 6 meters, 0.25 mm ID,
          0.25 μm film thickness
Temp: initial temp. 60° C
          ramp to 65° C at 25 sec
          ramp to 80° C at 50 sec
          ramp to 100° C at 60 sec
          ramp to 225° C at 80 sec
Carrier: Hydrogen
Flow: 4 mL/min.
Velosity: 100 cm/sec.
Split: 50:1
Detector: FID

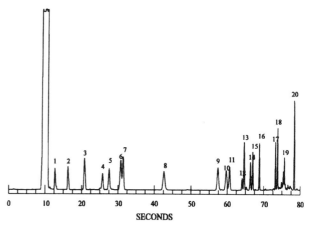

Fig. 1.6. Comparison of orange oil separations; (a) a conventional separation (b) a fast separation on a Flash-GC instrument. Reprinted with permission of Thermedics Detection.

greatest plate count—the "best" column in the world—and since column efficiency increases with column length, this has led to a competition to make the longest column. Currently, the record for the longest continuous column is held by Chrompack International [11] who made a 1300-m fused silica column (the largest size that would fit inside a commercial GC oven). It had a plate number of 1.2 million which was smaller than predicted, due in part to limits in the operational conditions.

Recently, a more efficient column was made by connecting nine 50-m columns into a single one of 450 m total length [12]. While much shorter than the Chrompack column, its efficiency was nearly 100% of theoretical, and it was calculated to have a plate number of 1.3 million and found capable of separating 970 components in a gasoline sample.

Because GC is excellent for quantitative analysis, it has found wide use for many different applications. Sensitive, quantitative detectors provide fast, accurate analyses, and at a relatively low cost. A pesticide separation illustrating the high speed, sensitivity, and selectivity of GC is shown in Figure 1.7.

GC has replaced distillation as the preferred method for separating volatile materials. In both techniques, temperature is a major variable, but gas chromatographic separations are also dependent upon the chemical nature (polarity) of the stationary phase. This additional variable makes GC more powerful. In addition, the fact that solute concentrations are very dilute in GC columns eliminates the possibility of azeotropes, which often plagued distillation separations.

Both methods are limited to volatile samples. A practical upper temperature limit for GC operation is about 380°C so samples need to have an appreciable vapor pressure (60 torr or greater) at that temperature. Solutes usually do not exceed boiling points of 500°C and molecular weights of

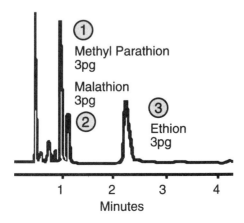

Fig. 1.7. Pesticide separation showing both high speed and low detectivity.

1,000 Daltons. This major limitation of GC is listed below along with other disadvantages of GC.

Disadvantages of Gas Chromatography

- Limited to volatile samples
- Not suitable for thermally labile samples
- Fairly difficult for large, preparative samples
- Requires spectroscopy, usually mass spectroscopy, for confirmation of peak identity

In summary: for the separation of volatile materials, GC is usually the method of choice due to its speed, high resolution capability, and ease of use.

INSTRUMENTATION

Figure 1.8 shows the basic parts of a simple gas chromatograph—carrier gas, flow controller, injector, column, detector, and data system. More detail is given in the next chapter.

The heart of the chromatograph is the column; the first ones were metal tubes packed with inert supports on which stationary liquids were coated. Today, the most popular columns are made of fused silica and are open tubes (OT) with capillary dimensions. The stationary liquid phase is coated on the inside surface of the capillary wall. The two types are shown in

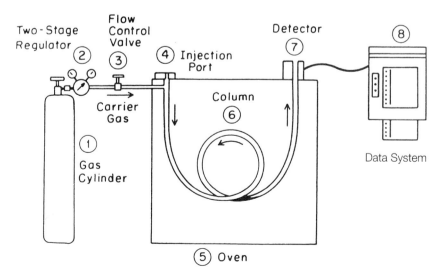

Fig. 1.8. Schematic of a typical gas chromatograph.

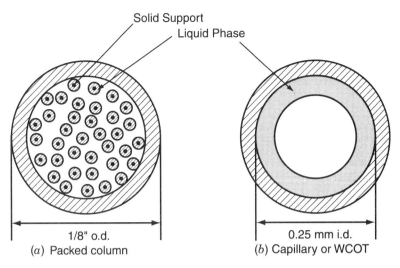

Fig. 1.9.   Schematic representation of (*a*) packed column and (*b*) open tubular column.

Figure 1.9 and each type is treated in a separate chapter—packed columns in Chapter 5 and capillary columns in Chapter 6.

REFERENCES

1. McNair, H., *LC-GC*, **10**, 239 (1992).
2. Ramsey, W., *Proc. Roy. Soc.* **A76**, 111 (1905).
3. Tswett, M., *Ber. deut. botan. Ges.*, **24**, 316 and 384 (1906).
4. Strain, H. H., and Sherma, J., *J. Chem. Educ.*, **44**, 238 (1967).
5. James, A. T., and Martin, A. J. P., *Biochem. J.*, **50**, 679 (1952).
6. Martin, A. J. P., and Synge, R. L. M., *Biochem. J.*, **35**, 1358 (1941).
7. Ettre, L. S., *Anal. Chem.*, **43**, [14], 20A–31A (1971)
8. Ettre, L. S., *LC-GC*, **8**, 716–724 (1990).
9. Ettre, L. S., *J. Chromatogr.* **112**, 1–26 (1975).
10. Ettre, L. S., *Pure & Appl. Chem.*, **65**, 819–872 (1993). See also, Ettre, L. S., *LC-GC*, **11,** 502 (1993).
11. de Zeeuw, J., Chrompack International B. V., Middleburg, the Netherlands, personal communication, 1996.
12. Berger, T. A., *Chromatographia*, **42**, 63 (1996).

2. Instrument Overview

Instrumentation in gas chromatography has continually evolved since the introduction of the first commercial systems in 1954. The basic components of a typical, *modern* gas chromatographic system are discussed individually in this chapter.

Figure 2.1 shows schematically a gas chromatographic system. The components which will be discussed include: (1) carrier gas; (2) flow control; (3) sample inlet and sampling devices; (4) columns; (5) controlled temperature zones (ovens); (6) detectors; and (7) data systems.

In summary, a gas chromatograph functions as follows. An inert carrier gas (like helium) flows continuously from a large gas cylinder through the injection port, the column, and the detector. The flow rate of the carrier gas is carefully controlled to ensure reproducible retention times and to minimize detector drift and noise. The sample is injected (usually with a microsyringe) into the heated injection port where it is vaporized and carried into the column, typically a capillary column 15 to 30 m long, coated on the inside with a thin (0.2 μm) film of high boiling liquid (the stationary phase). The sample partitions between the mobile and stationary phases, and is separated into individual components based on relative solubility in the liquid phase and relative vapor pressures.

After the column, the carrier gas and sample pass through a detector. This device measures the quantity of the sample, and generates an electrical signal. This signal goes to a data system/integrator which generates a chromatogram (the written record of analysis). In most cases the data-handling

14

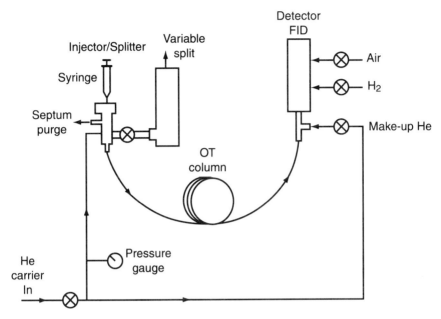

Fig. 2.1. Schematic of a typical gas chromatograph.

system automatically integrates the peak area, performs calculations and prints out a report with quantitative results and retention times. Each of these seven components will be discussed in greater detail.

CARRIER GAS

The main purpose of the carrier gas is to carry the sample through the column. It is the *mobile phase* and it is inert and does not interact chemically with the sample.

A secondary purpose is to provide a suitable matrix for the detector to measure the sample components. Below are the carrier gases preferred for various detectors:

CARRIER GASES AND DETECTORS

Detector	Carrier Gas
Thermal conductivity	Helium
Flame ionization	Helium or nitrogen
Electron capture	Very dry nitrogen or Argon, 5% methane

For the thermal conductivity detector, helium is the most popular. While hydrogen is commonly used in some parts of the world (where helium is very expensive), it is not recommended because of the potential for fire and explosions. With the flame ionization detector, either nitrogen or helium may be used. Nitrogen provides slightly more sensitivity, but a slower analysis than helium. For the electron capture detector, very dry, oxygen-free nitrogen, or a mixture of argon with 5% methane is recommended.

Purity

It is important that the carrier gas be of high purity because impurities such as oxygen and water can chemically attack the liquid phase in the column and destroy it. Polyester, polyglycol and polyamide columns are particularly susceptible. Trace amounts of water can also desorb other column contaminants and produce a high detector background or even "ghost peaks." Trace hydrocarbons in the carrier gas cause a high background with most ionization detectors and thus limit their detectability.

One way to obtain high purity carrier gas is to purchase high purity gas cylinders. The following list compares the purity and prices for helium available in the United States:

HELIUM SPECIFICATIONS AND PRICES

Quality	Purity	Price
Research grade	99.9999%	$280
Ultrapure	99.999%	$140
High purity	99.995%	$55

Prices are quoted for a cylinder containing 49 liters (water capacity) and rated at 2400 psi. Obviously, purchasing the carrier gas of adequate purity is not economically feasible for most laboratories.

The more common practice is to purchase the High Purity grade and purify it. Water and trace hydrocarbons can be easily removed by installing a 5A molecular sieve filter between the gas cylinder and the instrument. Drying tubes are commercially available, or they can be readily made by filling a 6-ft. by 1/4" column with GC grade 5A molecular sieve. In either case, after two gas cylinders have been used, the sieve should be regenerated by heating to 300°C for 3 hours with a slow flow of dry nitrogen. If home-made, the 6-ft. column can be coiled to fit easily into the chromatographic column oven for easy regeneration.

Oxygen is more difficult to remove and requires a special filter, such as a BTS catalyst from BASF, Ludwigshaven am Rhein, Oxisorb from Supelco, or Dow Gas Purifier from Alltech.

FLOW CONTROL AND MEASUREMENT

The measurement and control of carrier gas flow is essential for both column efficiency and for qualitative analysis. Column efficiency depends on the proper linear gas velocity which can be easily determined by changing the flow rate until the maximum plate number is achieved. Typical optimum values are: 75 to 90 mL/min for 1/4″ outside diameter (o.d.) packed columns; 25 mL/min for 1/8″ o.d. packed columns; and 0.75 mL/min for a 0.25 μm i.d. open tubular column. These values are merely guidelines; the optimum value for a given column should be determined experimentally.

For qualitative analysis, it is essential to have a constant and reproducible flow rate so that retention times can be reproduced. Comparison of retention times is the quickest and easiest technique for compound identification. Keep in mind that two or more compounds may have the same retention time, but no compound may have two different retention times. Thus, retention times are characteristic of a solute, but not unique. Obviously, good flow control is essential for this method of identification.

Controls

The first control in any flow system is a two-stage regulator connected to the carrier gas cylinder to reduce the tank pressure of 2,500 psig down to a useable level of 20–60 psig. It should include a safety valve and an inlet filter to prevent particulate matter from entering it. A stainless steel diaphragm is recommended to avoid any air leaks into the system. The first gauge indicates the pressure left in the gas cylinder. By turning the valve on the second stage, an increasing pressure will be delivered to the gas chromatograph and will be indicated on the second gauge. The second stage regulator does not work well at low pressures and it is recommended that a minimum of 20 psi be used.

For isothermal operation, constant pressure is sufficient to provide a constant flow rate, assuming that the column has a constant pressure drop. For simple, inexpensive gas chromatographs which run only isothermally, the second part of the flow control system may be a simple needle valve; this, however, is not sufficient for research systems.

In temperature programming, even when the inlet pressure is constant, the flow rate will decrease as the column temperature increases. As an example, at an inlet pressure of 24 psi and a flow rate of 22 mL/min (helium) at 50°C, the flow rate decreases to 10 mL/min at 200°C. This decrease is due to the increased viscosity of the carrier gas at higher temperatures. In all temperature-programmed instruments, and even in some better isothermal ones, a differential flow controller is used to assure a constant mass flow rate.

Sometimes, however, it is not desirable to control the flow rate with such a controller. For example, split and splitless sample injection both depend on a constant *pressure* for correct functioning. Constant pressure maintains the same flow rate through the column, independent of the opening and closing of the purge valve. Under these conditions, the carrier gas pressure can be increased electronically during a programmed run in order to maintain a constant flow. An electronic sensor is used to detect the (decreasing) flow rate and increase the pressure to the column, thus providing a constant flow rate by electronic pressure control (EPC).

Flow Measurement

The two most commonly used devices are a soap-bubble flowmeter and a digital electronic flow measuring device (Fig. 2.2). The soap-film flowmeter is merely a calibrated tube (usually a modified pipet or buret) through which the carrier gas flows. By squeezing a rubber bulb, a soap solution is raised into the path of the flowing gas. After several soap bubbles are allowed to wet the tube, one bubble is accurately timed through a defined volume with a stopwatch. From this measurement, the carrier gas flow rate in mL/min is easily calculated. Some electronic flow meters are based on the same principle, but the measurements are made with light beams. At a cost around $300, an electronic flow meter is faster and easier to use, and provides a three-digit readout of the flow rate.

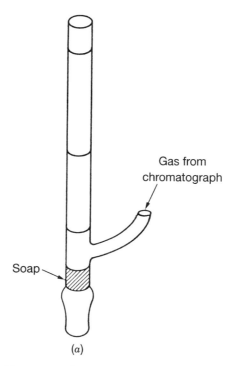

Gas from
chromatograph

Soap

(a)

Fig. 2.2. Flow meters: (a) soap film type (b) digital electronic type.

Another, more sophisticated electronic device uses a solid-state sensor coupled with a microprocessor to permit accurate flow measurements for a range of gases without using soap bubbles. A silicone-on-ceramic sensor can be used to measure flow rates of 0.1 to 500 mL/min for air, oxygen, nitrogen, helium, hydrogen, and 5% argon in methane. The cost for this device is about $500.

Very small flow rates such as those encountered in open tubular columns, cannot be measured reliably with these meters. The average linear flow velocity in OT columns, $\bar{u}$, can be calculated from equation 1:

$$\bar{u} = \frac{L}{t_M} \qquad (1)$$

where L is the length of the column (cm) and t_M is the retention time for a nonretained peak such as air or methane (seconds). Since the flame detector does not detect air, methane is usually used for this measurement, but the column conditions must be chosen (high enough temperature) so that it is not retained. Conversion of the linear velocity in cm/sec to flow

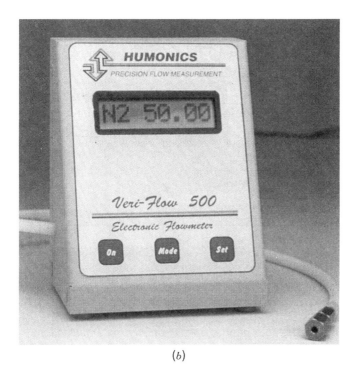

(*b*)

Fig. 2.2. *(Continued).*

rate (in mL/min) is achieved by multiplying by the cross-sectional area of the column (πr^2). See Appendix IV.

Compressibility of the Carrier Gas

Since the carrier gas entering a GC column is under pressure and the column outlet is usually at atmospheric pressure, the inlet pressure, p_i, is greater than the outlet pressure, p_O. Consequently, the gas is compressed at the inlet and expands as it passes through the column; the volumetric flow rate also increases from the head of the column to the outlet.

Usually the volumetric flow rate is measured at the outlet where it is at a maximum. To get the average flow rate, $\overline{F}_c$, the outlet flow must be multiplied by the so-called *compressibility factor, j:*

$$j = \frac{3}{2}\left[\frac{\left(\dfrac{p_i}{p_o}\right)^2 - 1}{\left(\dfrac{p_i}{p_o}\right)^3 - 1}\right] \tag{2}$$

and:

$$\overline{F}_c = j \times F_c \tag{3}$$

Some typical values of j are given in the Appendix VII.

If one calculates a retention volume from a retention time, the average flow rate should be used, and the resulting retention volume is called the *corrected* retention volume, V_R^0:

$$V_R^0 = j\, V_R = j\, t_R\, F_c \tag{4}$$

This term should not be confused with the *adjusted* retention volume to be presented in the next chapter.

SAMPLE INLETS AND SAMPLING DEVICES

The sample inlet should handle a wide variety of samples including gases, liquids, and solids, and permit them to be rapidly and quantitatively introduced into the carrier gas stream. Different column types require different types of sample inlets as indicated in the following list:

SAMPLE INLETS

Packed Column	Capillary Column
Flash vaporizer	Split
On-column	Splitless
	On-column

Ideally, the sample is injected instantaneously onto the column, but in practice this is impossible and a more realistic goal is to introduce it as a sharp symmetrical band. The difficulty keeping the sample sharp and narrow can be appreciated by considering the vaporization of a 1.0 microliter sample of benzene. Upon injection, the benzene vaporizes to 600 μL of vapor. In the case of a capillary column (at a flow rate of 1 mL/min), 36 seconds would be required to carry it onto the column. This would be so slow that an initial broad band would result and produce very poor column performance (low N). Clearly, sampling is a very important part of the chromatographic process and the size of the sample is critical.

There is no single optimum sample size. Some general guidelines are available, however. Table 2.1 lists typical sample sizes for three types of columns. For the best peak shape and maximum resolution, the smallest possible sample size should always be used.

The more components present in the sample, the larger the sample size may need to be. In most cases, the presence of other components will not affect the location and peak shape of a given solute. For trace work, and for preparative-scale work, it is often best to use large sample sizes even though they will "overload" the column. The major peaks may be badly distorted, but the desired (trace) peaks will be larger, making it possible to achieve the desired results.

Gas Sampling

Gas sampling methods require that the entire sample be in the gas phase under the conditions in use. Mixtures of gases and liquids pose special problems. If possible, mixtures should either be heated, to convert all components to gases, or pressurized, to convert all components to liquids. Unfortunately, this is not always possible.

TABLE 2.1 Sample Volumes for Different Column Types

Column Types	Sample Sizes (liquid)
Regular analytical packed: $\frac{1}{4}''$ o.d., 10% liquid	0.2–20 μl
High efficiency packed: $\frac{1}{8}''$ o.d., 3% liquid	0.01–2 μl*
Capillary (open tubular): 250 μm i.d., 0.2 μm film	0.01–3 μl*

* These sample sizes are often obtained by sample splitting techniques.

Gas-tight syringes and gas sampling valves are the most commonly used methods for gas sampling. The syringe is more flexible, less expensive and the most frequently used device. A gas-sampling valve on the hand gives better repeatability, requires less skill and can be more easily automated. Refer to Chapter 5 for more details on valves.

Liquid Sampling

Since liquids expand considerably when they vaporize, only small sample sizes are desirable, typically microliters. Syringes are almost the universal method for injection of liquids. The most commonly used sizes for liquids are 1, 5, and 10 microliters. In those situations where the liquid samples are heated (as in all types of vaporizing injectors) to allow rapid vaporization before passage into the column, care must be taken to avoid overheating that could result in thermal decomposition.

Solid Sampling

Solids are best handled by dissolving them in an appropriate solvent, and by using a syringe to inject the solution.

Syringes

Figure 2.3 shows a 10-microliter liquid syringe typically used for injecting one to five microliters of liquids or solutions. The stainless steel plunger fits tightly inside a precision barrel made of borosilicate glass. The needle, also stainless steel, is epoxied into the barrel. Other models have a removable needle that screws onto the end of the barrel. For smaller volumes, a 1-microliter syringe is also available. A 10-milliliter gas-tight syringe is used for injecting gaseous samples up to about 5 milliliters in size. A useful suggestion is to always use a syringe whose total sample volume is at least two times larger than the volume to be injected.

Using a Syringe

In filling a microliter syringe with liquid, it is desirable to exclude all air initially. This can be accomplished by repeatedly drawing liquid into the syringe and rapidly expelling it back into the liquid. Viscous liquids must be drawn into the syringe slowly; very fast expulsion of a viscous liquid

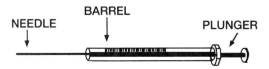

Fig. 2.3.   Microsyringe, $10\mu L$ volume.

could split the syringe. If too viscous, the sample can be diluted with an appropriate solvent.

Draw up more liquid into the syringe than you plan to inject. Hold the syringe vertically with the needle pointing up so any air still in the syringe will go to the top of the barrel. Depress the plunger until it reads the desired value; the excess air should have been expelled. Wipe off the needle with a tissue, and draw some air into the syringe now that the exact volume of liquid has been measured. This air will serve two purposes: first, it will often give a peak on the chromatogram, which can be used to measure t_M; second, the air prevents any liquid from being lost if the plunger is accidentally pushed.

To inject, use one hand to guide the needle into the septum and the other to provide force to pierce the septum and also to prevent the plunger from being blown out by the pressure in the GC. The latter point is important when large volumes are being injected (e.g., gas samples) or when the inlet pressure is extremely high. Under these conditions, if care is not exercised, the plunger will be blown out of the syringe.

Insert the needle rapidly through the septum and as far into the injection port as possible and depress the plunger, wait a second or two, then withdraw the needle (keeping the plunger depressed) as rapidly and smoothly as possible. Note that alternate procedures are often used with open tubular columns. Be careful; most injection ports are heated and you can easily burn yourself.

Between samples, the syringe must be cleaned. When high-boiling liquids are being used, it should be washed with a volatile solvent like methylene chloride or acetone. This can be done by repeatedly pulling the wash liquid into the syringe and expelling it. Finally, the plunger is removed and the syringe dried by pulling air through it with a vacuum pump (appropriately trapped) or a water aspirator. Pull the air in through the needle so dust cannot get into the barrel to clog it. Wipe the plunger with a tissue and reinsert. If the needle gets dulled, it can be sharpened on a small grindstone.

Autosamplers

Samples can be injected automatically with mechanical devices that are often placed on top of gas chromatographs. These autosamplers mimic the human injection process just described using syringes. After flushing with solvent, they draw up the required sample several times from a sealed vial and then inject a fixed volume into the standard GC inlet. Autosamplers consist of a tray which holds a large number of samples, standards, and wash solvents, all of which are rotated into position under the syringe as needed. They can run unattended and thus allow many samples to be run overnight. Autosamplers provide better precision than manual injection—typically 0.2% relative standard deviation (RSD).

Septa

Syringe injection is accomplished through a self-sealing septum, a polymeric silicone with high-temperature stability. Many types of septa are commercially available; some are composed of layers and some have a film of Teflon® on the column side. In selecting one, the properties that should be considered are: high temperature stability, amount of septum "bleed" (decomposition), size, lifetime, and cost.

COLUMNS

Figure 2.4 shows schematically a packed column in a longitudinal cross section. The column itself is usually made of stainless steel and is packed tightly with stationary phase on an inert solid support of diatomaceous earth coated with a thin film of liquid. The liquid phase typically constitutes 3, 5, or 10% by weight of the total stationary phase.

Packed columns are normally three, six, or twelve feet in length. The outside diameter is usually 1/4" or 1/8". Stainless steel is used most often, primarily because of its strength. Glass columns are more inert, and they are often used for trace pesticide and biomedical samples that might react with the more active stainless steel tubing.

Packed columns are easy to make and easy to use. A large variety of liquid phases is available. Because the columns are tightly packed with small particles, lengths over 20 feet are impractical, and only a modest number of plates is usually achieved (about 8,000 maximum). Packed columns are covered in detail in Chapter 5.

Capillary columns are simple chromatograpic columns, which are not filled with packing material. Instead, a thin film of liquid phase coats the inside wall of the 0.25 mm fused silica tubing. Such columns are properly

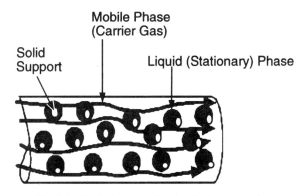

Fig. 2.4. Packed column, longitudinal cross section.

called "wall-coated open tubular" or simply WCOT columns. Since the tube is open, its resistance to flow is very low; therefore, long lengths, up to 100 meters, are possible. These long lengths permit very efficient separations of complex sample mixtures. Fused silica capillary columns are the most inert. Open tubular (OT) columns are covered in detail in Chapter 6.

Table 2.2 compares these two main types of column and lists their advantages, disadvantages, and some typical applications.

TEMPERATURE ZONES

The column is thermostated so that a good separation will occur in a reasonable amount of time. It is often necessary to maintain the column at a wide variety of temperatures, from ambient to 360°C. The control of temperature is one of the easiest and most effective ways to influence the separation. The column is fixed between a heated injection port and a heated detector, so it seems appropriate to discuss the temperature levels at which these components are operated.

Injection-port Temperature

The injection port should be hot enough to vaporize the sample rapidly so that no loss in efficiency results from the injection technique. On the other hand, the injection-port temperature must be low enough so that thermal decomposition or chemical rearrangement is avoided.

For flash vaporization injection, a general rule is to have the injection temperature about 50°C hotter than the boiling point of the sample. A practical test is to raise the temperature of the injection port. If the column

TABLE 2.2 Comparison of Packed and WCOT Columns

	$\frac{1}{8}''$ Packed	WCOT
Outside diameter	3.2 mm	0.40 mm
Inside diameter	2.2 mm	0.25 mm
d_f	$5\mu m$	$0.25 \ \mu m$
β	15–30	250
Column length	1–2 m	15–60 m
Flow	20 mL/min	1 mL/min
N_{tot}	4,000	180,000
H_{min}	0.5 mm	0.3 mm
Advantages	Lower cost	Higher efficiency
	Easier to make	Faster
	Easier to use	More inert
	Larger samples	Fewer columns needed
	Better for fixed gases	Better for complex mixtures

efficiency or peak shape improves, the injection-port temperature was too low. If the retention time, the peak area, or the shape changes drastically, the temperature may be too high and decomposition or rearrangement may have occurred. For on-column injection, the inlet temperature can be lower.

Column Temperature

The column temperature should be high enough so that sample components pass through it at a reasonable speed. It need *not* be higher than the boiling point of the sample; in fact is is usually preferable if the column temperature is considerably below the boiling point. If that seems illogical, remember that the column operates at a temperature where the sample is in the *vapor* state—it need not be in the *gas* state. In GC, the column temperature must be kept above the "dew point" of the sample, but not above its boiling point.

In Figure 2.5, a hydrocarbon sample is run on the same column at 75, 110, and 130°C. At 75°C the vapor pressures of the sample components are low and they move slowly through the column. Two isomers of octane are well resolved before the C-8 peak; however, the analysis time is very long, at 24 minutes.

At higher temperatures, the retention times decrease. At 110°C the C-12 peak is out in 8 minutes and by 130°C the analysis is complete in 4 minutes, but the resolution decreases. Notice that the octane isomers are no longer resolved at the higher temperature. Lower temperature means longer analysis times, but better resolution.

Isothermal vs. Programmed Temperature (PTGC)

Isothermal denotes a chromatographic analysis at one constant column temperature. Programmed temperature refers to a linear increase of column

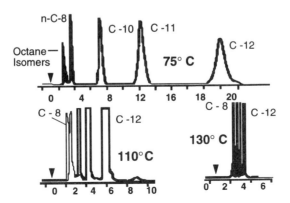

Fig. 2.5.   Effect of temperature on retention time.

temperature with time. Temperature programming is very useful for wide boiling sample mixtures and is very popular. Further details can be found in Chapter 9.

Detector Temperature

The detector temperature depends on the type of detector employed. As a general rule, however, the detector and its connections from the column exit must be hot enough to prevent condensation of the sample and/or liquid phase. If the temperature is too low and condensation occurs, peak broadening and even the total loss of peaks is possible.

The thermal conductivity detector temperature must be controlled to $\pm 0.1°C$ or better for baseline stability and maximum detectivity. Ionization detectors do not have this strict a requirement; their temperature must be maintained high enough to avoid condensation of the samples and also of the water or by-products formed in the ionization process. A reasonable minimum temperature for the flame ionization detector is 125°C.

DETECTORS

A detector senses the effluents from the column and provides a record of the chromatography in the form of a chromatogram. The detector signals are proportionate to the quantity of each solute (analyte) making possible quantitative analysis.

The most common detector is the flame ionization detector, FID. It has the desirable characteristics of high sensitivity, linearity, and detectivity and yet it is relatively simple and inexpensive. Other popular detectors are the thermal conductivity cell (TCD) and the electron capture detector (ECD). These and a few others are described in Chapter 7.

DATA SYSTEMS

Since OT columns produce fast peaks, the major requirement of a good data system is the ability to measure the GC signal with rapid sampling rates. Currently there is an array of hardware, made possible by advances in computer technology, that can easily perform this function. In general, there are two types of systems in common use—integrators and computers.

Microprocessor-based integrators are simply hard wired, dedicated micro processors which use an analog-to-digital (A-to-D) converter to produce both the chromatogram (analog signal) and a digital report for quantitative analysis. They basically need to calculate the start, apex, end, and area of each peak. Algorithms to perform these functions have been available for some time.

Most integrators perform area percent, height percent, internal standard, external standard, and normalization calculations. For nonlinear detectors, multiple standards can be injected, covering the peak area of interest, and software can perform a multilevel calibration. The operator then chooses an integrator calibration routine suitable for that particular detector output.

Many integrators provide BASIC programming, digital control of instrument parameters, and automated analysis, from injection to cleaning of the column and injection of the next sample. Almost all integrators provide an RS-232-C interface so the GC output is compatible with "in house" digital networks.

Personal computer-based systems have now successfully migrated to the chromatography laboratory. They provide easy means to handle single or multiple chromatographic systems and provide output to both local and remote terminals. Computers have greater flexibility in acquiring data, instrument control, data reduction, display and transfer to other devices. The increased memory, processing speed and flexible user interfaces make them more popular than dedicated integrators. Current computer-based systems rely primarily on an A-to-D card, which plugs into the PC main frame. Earlier versions used a separate stand-alone A-to-D box or were interfaced to stand-alone integrators. As costs for PCs decrease, their popularity will undoubtedly increase.

3. Basic Concepts and Terms

In Chapter 1, definitions and terms were presented to facilitate the description of the chromatographic system. In this chapter, additional terms are introduced and related to the basic theory of chromatography. Please refer to Table 1.1 in Chapter 1 for a listing of some of the symbols. Make special note of those that are recommended by the IUPAC; they are the ones used in this book.

This chapter continues with a presentation of the Rate Theory, which explains the processes by which solute peaks are broadened as they pass through the column. Rate theory treats the kinetic aspects of chromatography and provides guidelines for preparing efficient columns—columns that keep peak broadening to a minimum.

DEFINITIONS, TERMS, AND SYMBOLS

Distribution Constant

A thermodynamic equilibrium constant called the distribution constant, K_c was presented in Chapter 1 as the controlling parameter in determining how fast a given solute moves down a GC column. For a solute or analyte designated A,

$$K_c = \frac{[A]_S}{[A]_M} \tag{1}$$

where the brackets denote molar concentrations and the subscripts S and M refer to the stationary and mobile phases respectively. The larger the distribution constant, the more the solute sorbs in the stationary phase, and the longer it is retained on the column. Since this is an *equilibrium* constant, one would assume that chromatography is an equilibrium process. Clearly it is not, because the mobile gas phase is constantly moving solute molecules down the column. However, if the kinetics of mass transfer are fast, a chromatographic system will operate close to equilibrium and thus the distribution constant will be an adequate and useful descriptor.

Another assumption not usually stated is that the solutes do not interact with one another. That is, molecules of solute A pass through the column as though no other solutes were present. This assumption is reasonable because of the low concentrations present in the column and because the solutes are increasingly separated from each other as they pass through the column. If interactions do occur, the chromatographic results will deviate from those predicted by the theory; peak shapes and retention volumes may be affected.

Retention Factor

In making use of the distribution constant in chromatography, it is useful to break it down into two terms.

$$K_c = k \times \beta \tag{2}$$

β is the phase volume ratio and k is the retention factor.

$$\beta = \frac{V_M}{V_S} \tag{3}$$

For capillary columns whose film thickness, d_f, is known, β can be calculated by using equation 4,

$$\beta = \frac{(r_c - d_f)^2}{2 r_c d_f} \tag{4}$$

where r_c is the radius of the capillary column. If, as is usually the case, $r_c \gg d_f$, equation 4 reduces to:

$$\beta = \frac{r_c}{2 \, d_f} \tag{5}$$

For capillary columns, typical β-values are in the hundreds, about 10 times the value in packed columns for which β is not as easily evaluated. The phase volume ratio is a very useful parameter to know and can be helpful in selecting the proper column. Some typical values are given in Table 3.1.

The retention factor, k, is the ratio of the *amount* of solute (not the *concentraion* of solute) in the stationary phase to the *amount* in the mobile phase:

$$k = \frac{(W_A)_S}{(W_A)_M} \qquad (6)$$

The larger this value, the greater the amount of a given solute in the stationary phase, and hence, the longer it will be retained on the column. In that sense, retention factor measures the extent to which a solute is retained. As such, it is just as valuable a parameter as the distribution constant, and it is one that can be easily evaluated from the chromatogram.

To arrive at a useful working definition, equation 2 is rearranged and equation 3 is substituted into it, yielding:

$$k = \frac{K_c}{\beta} = \frac{K_c V_S}{V_M} \qquad (7)$$

Recalling the basic chromatographic equation introduced in Chapter 1,

$$V_R = V_M + K_c V_S \qquad (8)$$

TABLE 3.1   Phase Volume Ratios (β) for Some Typical Columns[a]

Column	Type[b]	I. D. (mm)	Length (m)	Film[c] Thickness (μm)	V_G (mL)	β	H (mm)	k^d
A	PC	2.16	2	10%	2.94	12	0.549	10.375
B	PC	2.16	2	5%	2.94	26	0.500	4.789
C	SCOT	0.50	15	—	2.75	20	0.950	6.225
D	WCOT	0.10	30	0.10	0.24	249	0.063	0.500
E	WCOT	0.10	30	0.25	0.23	99	0.081	1.258
F	WCOT	0.25	30	0.25	1.47	249	0.156	0.500
G	WCOT	0.32	30	0.32	2.40	249	0.200	0.500
H	WCOT	0.32	30	0.50	2.40	159	0.228	0.783
I	WCOT	0.32	30	1.00	2.38	79	0.294	1.576
J	WCOT	0.32	30	5.00	2.26	15	0.435	8.300
K	WCOT	0.53	30	1.00	6.57	132	0.426	0.943
L	WCOT	0.53	30	5.00	6.37	26	0.683	4.789

[a] Taken from Ref. 1. Reprinted with permission of the author.
[b] Type: PC = Packed Column
   SCOT = Support-coated Open Tubular
   WCOT = Wall-coated Open Tubular
[c] For packed columns: liquid stationary phase loading in weight percent.
[d] Relative values based on column G having $k = 0.5$.

and rearranging it produces a new term, V_R', the *adjusted* retention volume.

$$V_R - V_M = V_R' = K_C V_S \tag{9}$$

It is the *adjusted retention volume* which is directly proportional to the thermodynamic distribution constant and therefore the parameter often used in theoretical equations. In essence it is the retention time measured from the nonretained peak (air or methane) as was shown in Figure 1.5.

Rearranging equation 9 and substituting it into equation 7 yields the useful working definition of k:

$$k = \frac{V_R'}{V_M} = \left(\frac{V_R}{V_M}\right) - 1 \tag{10}$$

Since both retention volumes, V_R' and V_M, can be measured directly from a chromatogram, it is easy to determine the retention factor for any solute as illustrated in Figure 3.1. Relative values of k are included in Table 3.1 to aid in the comparison of the column types tabulated there.

Note that the more a solute is retained by the stationary phase, the larger is the retention volume and the larger is the retention factor. Thus, even though the distribution constant may not be known for a given solute, the retention factor is readily measured from the chromatogram, and it can be used instead of the distribution constant to measure the relative extent of sorption by a solute. However, if β is known (as is usually the case for OT columns), the distribution constant can be calculated from equation 2.

Because the definition of the *adjusted* retention volume was given above, and a related definition of the *corrected* retention volume was given in Chapter 2 (equation 4), we ought to make sure that these two are not confused with one another. Each has its own particular definition: the adjusted retention volume, V_R' is the retention volume excluding the void volume (measured from the methane or air peak) as shown in equation 9; the corrected retention volume, V_R^0, is the value correcting for the compres-

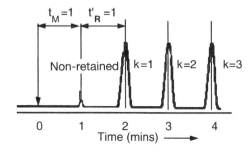

Fig. 3.1. Illustration of retention factor, *k*.

sibility of the carrier gas and based on the *average* flow rate. There is still another retention volume representing the value that is both adjusted *and* corrected; it is called the *net* retention volume, V_N:

$$V_N = j (V_R - V_M) = j V_R' = V_R^O - V_M^O \tag{11}$$

Consequently, for GC, equation 9 should more appropriately be written as:

$$V_N = K_C V_S \tag{12}$$

Depending on the particular point they are making, gas chromatographers feel free to substitute the adjusted retention volume in situations where they should be using the net retention volume. In LC, there is no significant compressibility of the mobile phase and the two values can be used interchangeably.

Retardation Factor

Another way to express the retention behavior of a solute is to compare its velocity through the column, μ, with the average[a] velocity of the mobile gas phase, $\bar{u}$:

$$\frac{\mu}{\bar{u}} = R \tag{13}$$

The new parameter defined by equation 13 is called the retardation factor, R. While it is not too widely used, it too can be calculated directly from chromatographic data, and it bears an interesting relationship to k.

To arrive at a computational definition, the solute velocity can be calculated by dividing the length of the column, L, by the retention time of a given solute,

$$\mu = \frac{L}{t_R} \tag{14}$$

where L is in cm or mm and the retention time is in seconds. Similarly, the average linear gas velocity is calculated from the retention time for a nonretained peak like air:

$$\bar{u} = \frac{L}{t_M} \tag{15}$$

---

[a] Remember from Chapter 2 that the linear velocity of the mobile phase varies through the column due to the compressibility of the carrier gas, so the value used in equation 12 is the *average* linear velocity, usually designated as $\bar{u}$.

Combining equations 10, 13, and 14 yields the computational definition of the retardation factor:

$$R = \frac{V_M}{V_R} \tag{16}$$

Because both of these volumes can be obtained from a chromatogram, the retardation factor is easily evaluated, as was the case for the retention factor.

Note that R and k are inversely related. To arrive at the exact relationship, equation 16 is substituted into equation 8, yielding:

$$R = \frac{1}{(1 + k)} \tag{17}$$

The retardation factor measures the extent to which a solute is retarded in its passage through the column, or the fractional rate at which a solute is moving. Its value will always be equal to, or less than, one.

It also represents the fraction of solute in the mobile phase at any given time and, alternatively, the fraction of time the average solute spends in the mobile phase. For example, a typical solute, A, might have a retention factor of 5, which means that it is retained 5 times longer than a non-retained peak. Its retardation factor, $1/(1 + k)$, is 1/6 or 0.167. This means that as the solute passed through the column, 16.7% of it was in the mobile phase and 84.3% was in the stationary phase at any instant. For another solute, B, with a retention factor of 9, the relative percentages are 10% in the mobile phase and 90% in the stationary phase. Clearly, the solute with the greater affinity for sorbing in the stationary phase, B in our example, spends a greater percentage of its time in the stationary phase, 90% versus 84.3% for A.

The retardation factor can also be used to explain how on-column injections work. When B is injected on-column, 90% of it sorbs into the stationary phase and only 10% goes into the vapor state. These numbers show that it is not necessary to "evaporate" all of the injected material; in fact, most of the solute goes directly into the stationary phase. Similarly, in Chapter 9, R will aid in our understanding of programmed temperature GC.

The retardation factor just defined for column chromatography is similar to the R_F factor in thin-layer chromatography, permitting liquid chromatographers to use these two parameters to compare TLC and HPLC data. And finally, it may be helpful in understanding the meaning of retention factor to note that the concept is similar in principle to the *fraction extracted* concept in liquid-liquid extraction.

Peak Shapes

We have noted that individual solute molecules act independently of one another during the chromatographic process. As a result, they produce a

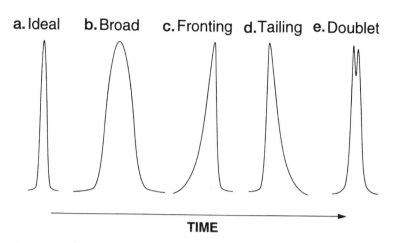

a. Ideal   b. Broad   c. Fronting d. Tailing e. Doublet

TIME

Fig. 3.2.   Peak shapes: (*a*) ideal, (*b*) broad, (*c*) fronting, (*d*) tailing, (*e*) doublet.

randomized aggregation of retention times after repeated sorptions and desorptions. The result for a given solute is a distribution, or peak, whose shape can be approximated as being *normal* or *Gaussian*. It is the peak shape that represents the ideal, and it is shown in all figures in the book except for those real chromatograms whose peaks are not ideal.

Nonsymmetrical peaks usually indicate that some undesirable interaction has taken place during the chromatographic process. Figure 3.2 shows some shapes that sometimes occur in actual samples. Broad peaks like (*b*) in Figure 3.2 are more common in packed columns and usually indicate that the kinetics of mass transfer are too slow (see The Rate Theory in this chapter). Sometimes, as in some packed column GSC applications (see Chapter 5), little can be done to improve the situation. However, it is the chromatographer's goal to make the peaks as narrow as possible in order to achieve the best separations.

Asymmetric peaks can be classified as tailing or fronting depending on the location of the asymmetry. The extent of asymmetry is defined as the tailing factor (TF) (Fig. 3.3).

$$TF = \frac{b}{a} \qquad (18)$$

Both *a* and *b* are measured at 10% of the peak height as shown.[b] As can be seen from the equation, a tailing peak will have a TF greater than one. The opposite symmetry, fronting, will yield a TF less than one. While the

[b] This is a commonly used definition, but unfortunately the USP definition is different. The latter definition of tailing is measured at 5% of the peak height and is: T = (a + b)/2a.

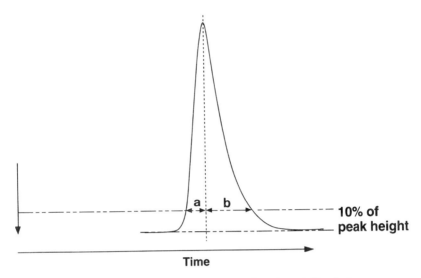

Fig. 3.3. Figure used to define asymmetric ratio or tailing factor.

definition was designed to provide a measure of the extent of tailing and is so named, it also measures fronting.

The doublet peak, like (e) in Figure 3.2, can represent a pair of solutes that are not adequately separated, another challenge for the chromatographer. Repeatability of a doublet peak should be verified because such a peak shape can also result from faulty injection technique, too much sample, or degraded columns (see Chapter 11).

For theoretical discussions in this chapter, ideal Gaussian peak shape will be assumed. The characteristics of a Gaussian shape are well known; Figure 3.4 shows an ideal chromatographic peak. The inflection points occur at 0.607 of the peak height and tangents to these points produce a triangle with a base width, w_b, equal to four standard deviations, 4σ, and a width at half height, w_h of 2.354σ. The width of the peak is 2σ at the inflection point (60.7% of the height). These characteristics are used in the definitions of some parameters, including the plate number.

Plate Number

To describe the efficiency of a chromatographic column, we need a measure of the peak width, but one that is relative to the retention time of the peak because width increases with retention time as we have noted before. Figure 3.5 illustrates this broadening phenomenon that is a natural consequence of the chromatographic process.

The most common measure of the efficiency of a chromatographic system is the plate number, N:

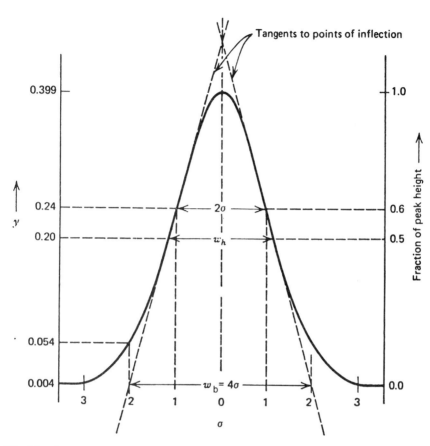

Fig. 3.4. A normal distribution. The inflection point occurs at 0.607 of the peak height. The quantity w_h is the width at 0.500 of the peak height (half-height) and corresponds to 2.354 σ. The quantity w_b is the base width and corresponds to 4σ as indicated. From Miller, J. M., *Chromatography: Concepts and Contrasts,* John Wiley & Sons, Inc., New York, 1987, p. 13. Reproduced courtesy of John Wiley & Sons, Inc.

$$N = \left(\frac{t_R}{\sigma}\right)^2 = 16 \left(\frac{t_R}{w_b}\right)^2 = 5.54 \left(\frac{t_R}{w_h}\right)^2 \qquad (19)$$

Figure 3.6 shows the measurements needed to make this calculation. Different terms arise because the measurement of σ can be made at different heights on the peak. At the base of the peak, w_b is 4σ, so the numerical constant is 4^2 or 16. At half height, w_h is 2.354σ and the constant becomes 5.54 (refer to Fig. 3.4).

Independent of the symbols used, both the numerator and the denominator must be given in the same units, and, therefore, N is unitless. Typically both the retention time and the peak width are measured as distances on

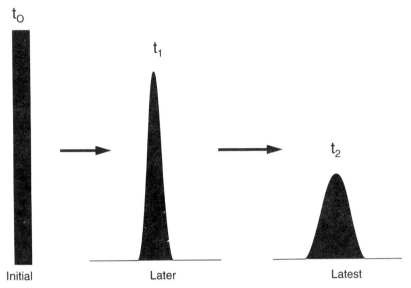

Fig. 3.5. Band broadening.

the chromatographic chart. Alternatively, both could be in either volume units or time units. No matter which calculation is made, a large value for N indicates an efficient column which is highly desirable.

For a chromatogram containing many peaks, the values of N for individual peaks may vary (they should increase slightly with retention time)

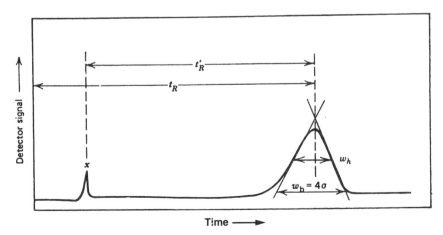

Fig. 3.6. Figure used to define plate number, N. The peak at x represents a nonretained component like air or methane. From Miller, J. M., *Chromatography: Concepts and Contrasts,* John Wiley & Sons, Inc., New York, 1987, p. 15. Reproduced courtesy of John Wiley & Sons, Inc.

depending on the accuracy with which the measurements are made. It is common practice, however, to assign a value to a particular column based on only one measurement even though an average value would be better.

Plate Height

A related parameter which expresses the efficiency of a column is the plate height, H,

$$H = \frac{L}{N} \tag{20}$$

where L is the column length. H has the units of length and is better than N for comparing efficiencies of columns of differing length. It is also called the *Height Equivalent to One Theoretical Plate* (HETP), a term which carried over from distillation terminology. Further discussion of H can be found later in this chapter. A good column will have a large N and a small H.

Resolution

Another measure of the efficiency of a column is resolution, R_S. As in other analytical techniques, the term resolution is used to express the degree to which adjacent peaks are separated. For chromatography, the definition is,

$$R_S = \frac{(t_R)_B - (t_R)_A}{\dfrac{(w_b)_A + (w_b)_B}{2}} = \frac{2d}{(w_b)_A + (w_b)_B} \tag{21}$$

where d is the distance between the peak maxima for two solutes, A and B. Figure 3.7 illustrates the way in which resolution is calculated. Tangents are drawn to the inflection points in order to determine the widths of the peaks at their bases. Normally, adjacent peaks of equal area will have the same peak widths, and $(w_b)_A$ will equal $(w_b)_B$. Therefore, equation 21 is reduced to:

$$R_S = \frac{d}{w_b} \tag{22}$$

In Figure 3.7, the tangents are just touching so $d = w_b$ and $R_S = 1.0$. The larger the value of resolution, the better the separation; complete, baseline separation requires a resolution of 1.5.

Strictly speaking, equations 21 and 22 are valid only when the heights of the two peaks are the same, as is shown in Figure 3.7. For other ratios of peak heights, the paper by Snyder [2] should be consulted for computer drawn examples.

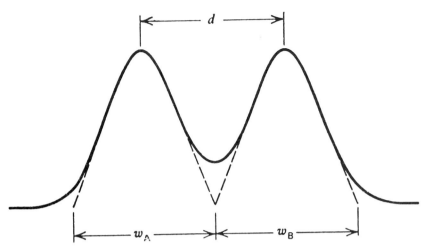

Fig. 3.7. Two nearly resolved peaks illustrating the definition of resolution, R_s. From Miller, J. M., *Chromatography: Concepts and Contrasts,* John Wiley & Sons, Inc., New York, 1987, p. 18. Reproduced courtesy of John Wiley & Sons, Inc.

Table 3.2 contains a summary of the most important chromatographic definitions and equations, and a complete list of symbols and acronyms is included in Appendix I.

THE RATE THEORY

The earliest attempts to explain chromatographic band broadening were based on an equilibrium model which came to be known as the Plate Theory. While it was of some value, it did not deal with the nonequilibrium conditions that actually exist in the column and did not address the causes of band broadening. However, an alternative approach describing the kinetic factors was soon presented; it became known as the Rate Theory.

The Original van Deemter Equation

The most influential paper using the kinetic approach was published by van Deemter, Klinkenberg, and Zuiderweg in 1956 [3]. It identified three effects that contribute to band broadening in packed columns; eddy diffusion (the A-term), longitudinal molecular diffusion (the B-term), and mass transfer in the stationary liquid phase (the C-term). The broadening was expressed in terms of the plate height, H, as a function of the average linear gas velocity, $\bar{u}$. In its simple form, the *van Deemter Equation* is:

TABLE 3.2   Some Important Chromatographic Equations and Definitions

1. $(K_C)_A = \dfrac{[A]_S}{[A]_M}$

2. $K_c = k\,\beta$

3. $\beta = \dfrac{V_M}{V_S}$

4. $\alpha = \dfrac{K_B}{K_A} = \dfrac{(V'_R)_B}{(V'_R)_A}$

5. $V_R = V_M + K_C V_S$

6. $V_N = K_C V_S$

7. $k = \dfrac{(W_A)_S}{(W_A)_M} = \dfrac{V'_R}{V_M} = \left(\dfrac{V_R}{V_M}\right) - 1 = \dfrac{1-R}{R} = \left(\dfrac{1}{R}\right) - 1$

8. $R = \dfrac{V_M}{V_R} = \dfrac{\mu}{u}$

   $\quad = \dfrac{V_M}{V_M + K_C V_S} = \dfrac{1}{1+k}$

9. $V_R = V_M(1+k) = \dfrac{L}{u}(1+k) = n(1+k)\dfrac{H}{u}$

10. $(1-R) = \dfrac{k}{k+1}$

11. $R(1-R) = \dfrac{k}{(k+1)^2}$

12. $N = 16\left(\dfrac{t_R}{w_b}\right)^2 = \left(\dfrac{t_R}{\sigma}\right)^2 = 5.54\left(\dfrac{t_R}{w_h}\right)^2$

13. $H = \dfrac{L}{n}$

14. $R_S = \dfrac{2d}{(w_b)_A + (w_b)_B}$

$$H = A + \frac{B}{u} + C\bar{u} \qquad (23)$$

Since plate height is inversely proportional to plate number, a small value indicates a narrow peak—the desirable condition. Thus, each of the three constants, *A, B,* and *C* should be minimized in order to maximize column efficiency.

The Golay Equation

Since open tubular or capillary columns do not have any packing, their rate equation does not have an *A*-term. This conclusion was pointed out by Golay [4], who also proposed a new term to deal with the diffusion process in the gas phase of open tubular columns. His equation had two *C*-terms, one for mass transfer in the stationary phase, C_S (similar to van

Deemter), and one for mass transfer in the mobile phase, C_M. The simple Golay equation is:

$$H = \frac{B}{u} + (C_S + C_M)\bar{u} \qquad (24)$$

The B-term of equation 24 accounts for the well-known molecular diffusion. The equation governing molecular diffusion is,

$$B = 2D_G \qquad (25)$$

where D_G is the diffusion coefficient for the solute in the carrier gas. Figure 3.8 illustrates how a zone of molecules diffuses from the region of high concentration to that of lower concentration with time. The equation tells us that a small value for the diffusion coefficient is desirable so that diffusion is minimized, yielding a small value for B and for H. In general, a low diffusion coefficient can be achieved by using carrier gases with larger molecular weights like nitrogen or argon. In the Golay equation (equation 24), this term is divided by the linear velocity, so a large velocity or flow rate will also minimize the contribution of the B-term to the overall peak broadening. That is, a high velocity will decrease the time a solute spends in the column and thus decrease the time available for molecular diffusion.

The C-terms in the Golay equation relate to mass transfer of the solute, either in the stationary phase or in the mobile phase. Ideally, fast solute

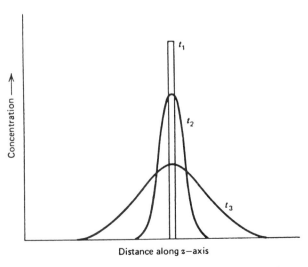

Fig. 3.8. Band broadening due to molecular diffusion. Three times are shown: $t_3 > t_2 > t_1$. From Miller, J. M., *Chromatography: Concepts and Contrasts,* John Wiley & Sons, Inc., New York, 1987, p. 31. Reproduced courtesy of John Wiley & Sons, Inc.

sorption and desorption will keep the solute molecules close together and keep the band broadening to a mimimum.

Mass transfer in the stationary phase can be described by reference to Figure 3.9. In both parts of the figure, the upper peak represents the distribution of a solute in the mobile phase and the lower peak the distribution in the stationary phase. A distribution constant of 2 is used in this example so the lower peak has twice the area of the upper one. At equilibrium, the solute achieves relative distributions like those shown in part (*a*), but an instant later the mobile gas moves the upper curve downstream giving rise to the situation shown in (*b*). The solute molecules in the stationary phase are stationary; the solute molecules in the gas phase have moved ahead of those in the stationary phase thus broadening the overall zone of molecules. The solute molecules which have moved ahead must now partition into the stationary phase and vice versa for those that are in the stationary phase, as shown by the arrows. The faster they can make this transfer, the less will be the band broadening.

The C_S-term in the Golay equation is,

$$C_S = \frac{2 \, k \, d_f^2}{3 \, (1 + k)^2 \, D_S} \tag{26}$$

where d_f is the average film thickness of the liquid stationary phase and D_S is the diffusion coefficient of the solute in the stationary phase. To minimize the contribution of this term, the film thickness should be small and the diffusion coefficient large. Rapid diffusion through thin films allows the solute molecules to stay closer together. Thin films can be achieved by coating small amounts of liquid on the capillary walls, but diffusion coefficients cannot usually be controlled except by selecting low viscosity stationary liquids.

Minimization of the C_S-term results when mass transfer into and out of the stationary liquid is as fast as possible. An analogy would be to consider a person jumping into and out of a swimming pool; if the water is shallow, the process can be done quickly; if it is deep, it cannot.

If the stationary phase is a solid, modifications in the C_S-term are necessary to relate it to the appropriate adsorption–desorption kinetics. Again, the faster the kinetics, the closer the process is to equilibrium, and the less is the band broadening.

The other part of the C_S-term is the ratio $k/(1 + k)^2$. Large values of k result from high solubilities in the stationary phase. This ratio is minimized at large values of k, but very little decrease occurs beyond a k-value of about 20. Since large values of retention factor result in long analysis times, little advantage is gained by k-values larger than 20.

Mass transfer in the mobile phase can be visualized by reference to Figure 3.10 which shows the profile of a solute zone as a consequence of

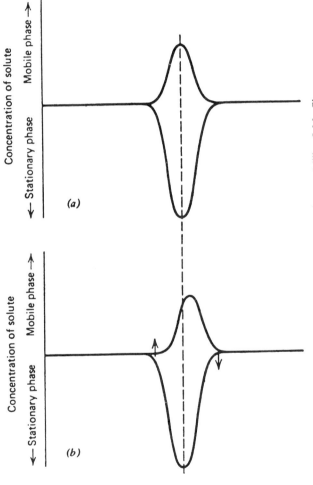

Fig. 3.9. Band broadening due to mass transfer. ($K_c = 2.0$). From Miller, J. M., *Chromatography: Concepts and Contrasts*, John Wiley & Sons, Inc., New York, 1987, p. 320. Reproduced courtesy of John Wiley & Sons, Inc.

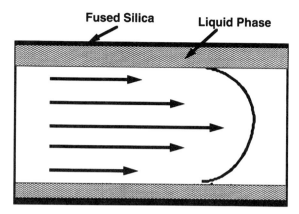

Fig. 3.10. Illustration of mass transfer in the mobile phase.

non-turbulent flow through a tube. Inadequate mixing (slow kinetics) in the gas phase can result in band broadening because the solute molecules in the center of the column move ahead of those at the wall. Small diameter columns minimize this broadening because the mass transfer distances are relatively small. Golay's equation for the C_M term is,

$$C_M = \frac{(1 + 6k + 11k^2) \, r_C^2}{24 \, (1 + k)^2 \, D_G} \tag{27}$$

where r_C is the radius of the column.

The relative importance of the two C-terms in the rate equation depends primarily on the film thickness and the column radius. Ettre [5] has published calculations for a few solutes on some typical 0.32 mm i.d. columns. A summary of his calculations is given in Table 3.3 showing that in thin films (0.25 μm) 95% of the total C-term is attributable to mass transfer in the mobile phase, (C_M), whereas for thick films (5.0 μm) it is only 31.5%.

TABLE 3.3 Relative Importance of Types of Mass Transfer[a]

Column	d_f (μm)	β	k	Relative Importance (%) C_M	C_S	Total of C-Terms Relative Magnitude
A	0.25	320	0.56	95.2	4.8	11
B	0.50	160	1.12	87.2	12.8	18
C	1.00	80	2.24	73.4	26.6	30
D	5.00	16	11.20	31.5	68.5	102

[a] Calculated data for *n*-undecane on 0.32 mm i.d. SE-30 column. Taken from Ref. 5. Reprinted with permission of the author.

Fig. 3.11. Illustration of eddy diffusion.

An extension of his calculations for other diameter columns shows that at smaller diameters (e.g., 0.25 mm), the C_M term is less dominate and for larger diameters (e.g., 0.53 mm) it is about three times as large, up to around 50%.

As a generalization, we can conclude that for thin films (<0.2 μm), the C-term is controlled by mass transfer in the mobile phase; for thick films (2–5.0 μm), it is controlled by mass transfer in the stationary phase; and for the intermediate films (0.2 to 2.0 μm) both factors need to be considered. For the larger "wide-bore" columns (see Chapter 6), the importance of mass transfer in the mobile phase is considerably greater.

Finally, we note that the C-terms are multiplied by the linear velocity in equation 24, so they are minimized at low velocities. Slow velocities allow time for the molecules to diffuse in and out of the liquid phase and to diffuse across the column in the mobile gas phase.

Mobile Phase Mass Transfer in Packed Columns

As originally proposed by van Deemter et al., the A-term dealt with eddy diffusion as shown in Figure 3.11. The diffusion paths of three molecules are shown in the figure. All three start at the same initial position, but they find differing paths through the packed bed and arrive at the end of the column, having traveled different distances. Because the flow rate of carrier gas is constant, they arrive at different times and are separated from each other. Thus, for a large number of molecules, the eddy diffusion process or the multi-path effect results in band broadening as shown.

The A-term in the van Deemter equation is,

$$A = 2\lambda d_p \tag{28}$$

where d_p is the diameter of the particles packed in the column and λ is a packing factor. To minimize A, small particles should be used and they should be tightly packed. In practice, the lower limit on the particle size is determined by the pressure drop across the column and the ability to pack uniformly very small particles. Mesh sizes around 100/120 are common.[c] Small ranges in size also promote better packing (minimal λ), so 100/120 is a better mesh range than say 80/120.

Since the original van Deemter equation did not include a C_M-term, an *extended* van Deemter equation that includes both the A-term and the C_M-term has been proposed [6]. A simplified version of the extended equation is:

$$H = A + \frac{B}{\bar{u}} + \frac{8\,k\,d_f^2\,\bar{u}}{\pi^2\,(1 + k)^2\,D_S} + \frac{\omega\,d_p^2\,\bar{u}}{D_G} \tag{29}$$

where ω is the obstruction factor for packed beds (a function of the solid support). This equation has found general acceptance although some others have been proposed and are discussed in the next section.

It should also be noted that the B-term in the original van Deemter equation included a tortuosity factor, γ, that also accounts for the nature of the packed bed. There is no such factor in the B-term for open tubular columns, of course.

Other Rate Equations

Additional modifications to the original van Deemter equation have been proposed by other workers. For example, one can argue that eddy diffusion (the A-term) is part of mobile phase mass transfer (the C_M-term) or is coupled with it. Giddings [7] has thoroughly discussed mass transfer and prefers a coupled term combining eddy diffusion and mass transfer to produce a new equation.

Others have defined rate equations that would serve both GC and LC [8]. An interesting discussion summarizing much of this work has been published by Hawkes [9]. His summary equation is in the same form as Golay's, but it is less specific. The references can be consulted for more information.

Van Deemter Plots

When the rate equation is plotted (H vs. $\bar{u}$), the so-called *van Deemter Plot* takes the shape of a nonsymmetrical hyperbola, shown in Figure 3.12. As one would expect from an equation in which one term is multiplied by velocity while another is divided by it, there is a minimum in the curve—an optimum velocity which provides the highest efficiency and smallest plate height.

---

[c] ASTM mesh sizes are given in the number of grids per inch in the sieve, so the larger the mesh size, the smaller the size of the particles that can pass through it. Mesh range 100/120 means that the particles are small enough to pass through a 100-mesh sieve but too large to pass a 120-mesh sieve.

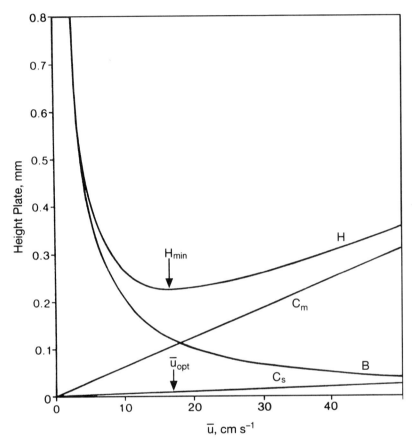

Fig. 3.12. Typical van Deemter plot. Courtesy of *Open-Tubular Column Gas Chromatography*, by Lee, Yang, and Bartle, John Wiley & Sons, Inc., 1984. Reprinted by permission of John Wiley & Sons, Inc.

It is logical to assume that chromatography would be carried out at the (optimum) velocity represented by the minimum in the curve since it yields the least peak broadening. However, if the velocity can be increased, the analysis time will be decreased. Consequently, chromatographers have devoted their time to manipulating the van Deemter equation to get the best performance for the shortest analysis time. By examining the relative importance of the individual terms to the overall equation in Figure 3.12, one sees that the upward slope as velocity is increased comes about from the increasing contribution of the *C*-terms. Therefore, most attention has been focused on minimizing them, a topic that will be covered shortly.

While the rate theory is a theoretical concept, it is a useful one in practice. It is common to obtain a van Deemter plot for one's column in order to evaluate it and the operating conditions. A solute is chosen and run isothermally at a variety of flow rates, being sure to allow sufficient time for pressure equilibration after each change. The plate number is evaluated from each chromatogram using equation 19 and then used to calculate the plate height (equation 20). The plate height values are plotted versus linear velocity (obtained by equation 15). The minimum velocity is noted as well as the slope of the curve at the higher velocities. Comparisons between columns will help in the selection of the best column. The van Deemter equation is seldom used to *calculate H.*

A Summary of the Rate Equations of van Deemter and Golay

Let us conclude this discussion by considering only two rate equations—one for open tubular columns and one for packed columns. The former is represented by the Golay equation:

$$H = \frac{2D_G}{\bar{u}} + \frac{2 \, k \, d_f^2 \, \bar{u}}{3 \, (1 + k)^2 \, D_S} + \frac{(1 + 6 \, k + 11 \, k^2) \, r_C^2 \, \bar{u}}{24 \, (1 + k)^2 \, D_G} \qquad (30)$$

and the latter by the extended van Deemter equation;

$$H = 2\lambda d_p + \frac{2\gamma \, D_G}{\bar{u}} + \frac{8k \, d_f^2 \, \bar{u}}{\pi^2 \, (1 + k)^2 \, D_S} + \frac{\omega d_p^2 \bar{u}}{D_G} \qquad (31)$$

Their value in improving chromatographic performance is summarized in the following section.

Practical Implications

Returning to our earlier suggestion that chromatographers look for ways to minimize both H and analysis time, let us compare the effect of carrier gas on the rate equation for a capillary column. One can choose to optimize the column efficiency (plate number) or the analysis time. For a given column, a higher molecular weight gas will generate more plates since the solute diffusivity is minimized (B-term). Nitrogen, having the higher molecular weight, shows a lower minimum H.

If one wishes to optimize the speed of analysis, however, it is better to choose a lighter carrier gas, like helium, or hydrogen. Referring to Figure 3.13, one sees that nitrogen has its minimum H at a linear gas velocity of 12 cm/sec. The minima for helium and hydrogen occur at about 20 and 40 cm/sec, respectively. If all gases were run at minimum H, nitrogen would generate about 15% more plates, but at an analysis time 3.3 times longer than hydrogen.

Finally, we must examine the slope of the curves beyond the minimum in Figure 3.13. We see that hydrogen, the lightest gas, has the smallest

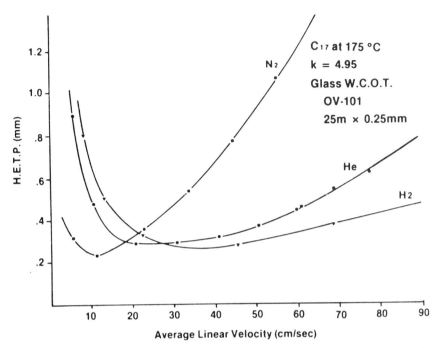

Fig. 3.13. Effect of carrier gas on van Deemter curve. (0.25 mm i.d. WCOT, $d_f = 0.4$ μm). Reprinted from Freeman, R. R., (ed.), *High Resolution Gas Chromatography,* Second edition, Hewlett-Packard Co., Wilmington, DE, 1981. Copyright 1981 Hewlett-Packard Company. Reproduced with permission.

slope. This means that with an increase in the hydrogen flow rate, a small loss in column efficiency can be offset by a large gain in the speed of analysis[d]. If one could choose the column length to optimize a given separation, the lighter carrier gases would provide the maximum plates per second, and thus the fastest analyses times.

As we have seen, the *C*-terms predominate at high velocities and column optimization is achieved by optimizing them. What factors contribute to an optimized *C*-term? Most important is the film thickness which should be small. Commercial columns are available with films of 0.1 μm although 0.25 μm films are more common. While thin films give high efficiencies and are good for high-boiling compounds, it should be remembered that they can accommodate only very small sample sizes.

Small diameter columns are desirable (small r_c in the C_M-term), especially if coated with thin films. The smallest commercial columns are 0.10 mm inside diameter (i.d.). Again, small sample sizes are required. Also, we

---

[d] The curves in Figure 3.12 are for isothermal operation. The advantages of light carrier gases is less pronounced in programmed temperature operation (PTGC).

have already noted that hydrogen is the preferred carrier gas for fast, efficient analyses; however, special care must be taken for its safe use.

For packed columns, thin films, narrow columns, and helium or hydrogen carrier gas are also desirable for efficient operation. The film thickness cannot be measured easily, and as an alternative, the percent by weight of liquid phase is usually given. The coverage is dependent on the surface area and density of the solid support; some equivalent loadings are given in Chapter 5 (Table 5.3). If too little stationary liquid is applied to the solid support, some of it will remain exposed and uncoated, usually resulting in undesirable adsorption and tailing. Typically columns with an o.d. of 1/8-inch are the smallest available commercially, although so-called micro-packed columns of 0.75 mm i.d. (1/6-inch o.d.) are available for a few phases.

In addition, the particles should be of small uniform size (e.g. 100–120 mesh) and be tightly and uniformly packed in the column. The inertness of the solid support is very important and is discussed in Chapter 5.

A REDEFINITION OF H

Now that we have related the plate height, H, to the important variables in the rate equation, it might be useful to look at it from another perspective. The concept of plate height originated in distillation theory where columns were described as containing plates or "theoretical" plates. Each plate occupied a certain space (height) in the distillation column, or, if there were no physical plates, each equilibrium stage was considered to be one theoretical plate. Thus, the plate height was the height of column occupied by one plate. In chromatography we have continued the use of these terms, and there is a plate theory that treats the chromatographic column as though it contained theoretical plates. However, our discussion of the rate theory has developed the concept of plate height as the extent of peak broadening for a solute as it passes through the column. Thus, a more appropriate term might be *column dispersivity* or *rate of band broadening*. In fact, another definition of H is,

$$H = \frac{\sigma^2}{L} \tag{32}$$

where σ^2 is the variance or square of the standard deviation representing the width of a peak, and L refers to the length (or distance) of movement of a solute.

A better understanding of the meaning of H can be obtained by combining several equations presented earlier, starting with the definition of H as equal to L/N. Substituting the definition of N (equation 19), we get:

$$H = \frac{L\sigma^2}{t_R^2} \tag{33}$$

Now we equate equations 32 and 33:

$$\frac{\sigma^2}{L} = \frac{L\sigma^2}{t_R^2} \tag{34}$$

and solve,

$$t_R = L \tag{35}$$

Consequently, the meaning of L for this situation (GC) is the retention time, and the concept of H for GC is best expressed as:

$$H = \frac{\sigma^2}{t_R} \tag{36}$$

or the variance (peak width) per unit time, or, rearranging:

$$\sigma = \sqrt{H\, t_R} \tag{37}$$

Equation 37 gives us a defintion of H which also provides the answer to the question about the extent of peak broadening during the chromatographic process: the peak width, expressed in σ, is proportional to the square root of the retention time. Thus, on a given column, a solute with a retention time twice that of another will have a peak width 1.4 times (square root of 2) the width of the other. Or, when using one solute to compare two columns that differ only in length, the width of the solute peak on the longer column will be the square root of the ratio of their lengths, times the width on the shorter column.

THE ACHIEVEMENT OF SEPARATION

We have seen how an analyte zone spreads or broadens as it passes through the chromatographic column. It might seem that this zone broadening is acting counter to our intention to separate solutes and could prevent chromatography from being effective. It is counterproductive, but it does not prevent us from achieving separations by chromatography.

Consider the simplified equation for resolution presented earlier in this chapter:

$$R_s = \frac{d}{w_b} \tag{38}$$

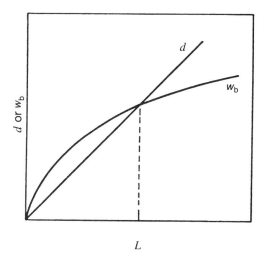

Fig. 3.14. The Achievement of Separation. Adapted from Giddings, J. C., *Dynamics of Chromatography,* Part. I, Marcel Dekker, New York, 1965, p. 33. Courtesy of Marcel Dekker, Inc. From Miller, J. M., *Chromatography: Concepts and Contrasts,* John Wiley & Sons, Inc., New York, 1987, p. 56. Reproduced courtesy of John Wiley & Sons, Inc.

While it is true that the peak width, here represented by w_b, increases as the square root of column length, L, the distance between two peaks, d, increases directly with L. Thus,

$$R_s \propto \frac{L}{\sqrt{L}} = \sqrt{L} \tag{39}$$

resolution is proportional to the square root of the column length.

This effect is shown graphically in Figure 3.14, where d and w_b are plotted against L. At some value of L, indicated by the dashed line, d becomes larger than w_b and separation is achieved. Our conclusion is that chromatography works, and as long as two solutes have *some* difference in their distribution constants, it must be possible to separate them if the column can be made long enough. That is, the chromatographic process is effective even though it produces peak broadening. In practice, of course, one seldom uses increasing column length as the only method for achieving separation.

REFERENCES

1. Ettre, L. S., *Chromatographia,* **18,** 477–488 (1984).
2. Snyder, L. R., *J. Chromatogr. Sci.,* **10,** 200 (1972).

3. van Deemter, J. J., Zuiderweg, F. J., and Klinkenberg, A., *Chem. Eng. Sci.,* **5,** 271 (1956).

4. Golay, M. J. E., *Gas Chromatography 1958,* D. H. Desty, Ed., Butterworths, London, 1958.

5. Ettre, L. C., *Chromatographia,* **17,** 553–559 (1983).

6. Jones, W. L., *Anal. Chem.,* **33,** 829 (1961).

7. Giddings, J. C. *Dynamics of Chromatography,* Part 1, Dekker, New York, 1965.

8. Knox, J. H., *Anal. Chem.,* **38,** 253 (1966).

9. Hawkes, S. J., *J. Chem. Educ.,* **60,** 393–398 (1983).

4. Stationary Phases

Of the two important decisions in setting up a gas chromatographic analysis, choosing the best column (usually the best stationary phase) is the more important. The other, selecting the column temperature, is less critical because the temperature can be easily programmed through a range of values to find the optimum value. (See Chapter 9.)

This chapter discusses the types of stationary phases, their classification, their applications, and the criteria used in selecting an appropriate liquid phase for a given separation. With packed columns, the choice of the stationary phase is critical, but it is less so for open tubular columns because of their higher efficiency. Individual chapters are devoted to each of the two column types, and this chapter is more relevant for packed columns (Chapter 5).

SELECTING A COLUMN

This section concerns the scientific basis for selecting a stationary phase, but first we must admit that there are other ways to select GC columns. The easiest and quickest is to ask someone who knows. That person may work in your laboratory or down the hall. If there is an experienced chromatographer nearby or otherwise accessible to you, you should not hesitate to ask.

There are also chromatography supply houses with extensive information, much of it already published—some in their catalogs. Increasingly, applications data are being made available in computerized form. Chrompack has produced a CD-ROM called CP-SCANVIEW and a diskette called CP-SCAN which contain over 1250 GC and LC applications. They have also put these data on their Web site on the Internet as has J&W Scientific, who has made its applications literature available on the Web. Ask them questions; give their applications chemist a call.

Another method is to make a search of the literature. GC is a mature science; it is highly probable that GC has already been applied to your type of sample as there are already over 100,000 GC publications. With ready access to Chemical Abstracts on-line, an experienced literature scientist should be able to come up with suggestions to help you.

A third choice is to go to the laboratory and make some trial runs. Some good columns and typical conditions are suggested in Table 4.1. With them, you can easily make a quick scouting run on your new sample.

CLASSIFICATION OF STATIONARY PHASES FOR GLC

In Chapter 1, it was noted that the stationary phase can be either a liquid or a solid. Liquids are more common and give rise to the subclassification known as gas-liquid chromatography, GLC. Solids and gas-solid chromatography, GSC, will be covered later in this chapter.

In order to use a liquid as the stationary phase in GC, some means must be found to hold the liquid in the column. For packed columns, the liquid is coated on a *solid support,* chosen for its high surface area and inertness. The coated support is dry-packed into the column as tightly as possible.

For open tubular (OT) or capillary columns, the liquid is coated on the inside of the capillary. To make it adhere better, the liquid phase is often extensively cross-linked and sometimes chemically bonded to the fused silica surface. See Chapter 6 for further details.

Liquid Phase Requirements

Hundreds of liquids have been used as stationary phases because the only requirements are a low vapor pressure, thermal stability, and if possible, a

TABLE 4.1 Recommended Columns for Scouting Runs

	Column*	
	Capillary	Packed
1. Stationary phase	DB-1	OV-101
2. Loading	0.25 μm	3% (w/w)
3. Column length	10 m	2 ft.
4. Column i.d.	0.25 mm	2 mm
5. Temperature program range (hold for 5 min. at max.)	60–320°C	100–300°C

* Packed column is glass; capillary column is fused silica.

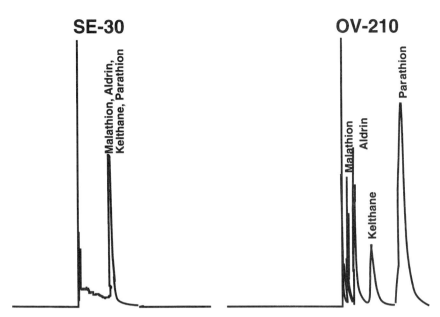

Fig. 4.1. Comparison of two liquid phases for an insecticide separation: (*a*) SE-30, a poor choice; *b*) OV-210®, a good choice. Both columns have the same efficiency, *N*.

low viscosity (for fast mass transfer). The large number of possible liquids has made the selection process complicated and some classification scheme is needed to simplify it.

Some examples will help to illustrate the effects of polarity on selectivity. To be effective as a stationary phase, the liquid chosen should interact with the components of the sample to be analyzed. The chemist's rule of thumb "like dissolves like" suggests that a polar liquid should be used to analyze polar analytes and a nonpolar liquid for nonpolar analytes. Figure 4.1 shows the separation of a pesticide mixture on two columns: a nonpolar SE-30[a] and a more polar OV-210[b]. Clearly, the selection of the proper stationary liquid is very important; in this case a *polar* column worked well for the *polar* pesticides. The nonpolar SE-30 is a good column (high efficiency) but it is not effective for this sample (small separation factor, α; see the next section).

In a comparison of two stationary phases which have extreme differences in polarity, the order of elution can be totally *reversed*. For example, Figure 4.2 shows the separation of four compounds that have similar boiling points on both a polar column, Carbowax® 20M[c], and a nonpolar column, SE-30 [1]. The elution order is reversed. The result of changing stationary-phase

[a] SE stands for Silicone Elastomer from the General Electric Co.
[b] OV designates the trade name for liquid phases from Ohio Valley Specialty Chemicals.
[c] Carbowax is the trade name of Union Carbide Corp. for their polyethyleneglycols.

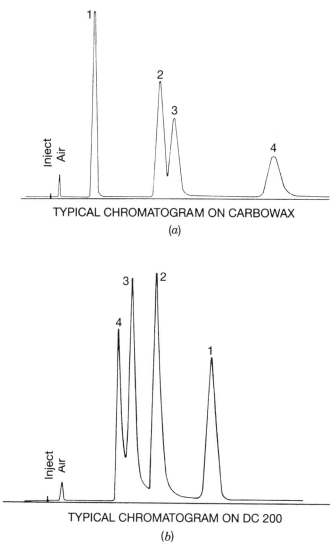

Fig. 4.2. Effect of stationary phase polarity on a 4-component separation: (*a*) Carbowax-20M® (polar); (*b*) DC-200 (nonpolar). Samples and their boiling points: (1) *n*-heptane (98); (2) tetrahydrofuran (64); (3) 2-butanone (80); (4) *n*-propanol (97). Reprinted with permission of the GOW-MAC Instrument Co., Bethlehem, PA, U.S.A.

polarity is not always so dramatic, but investigators should be aware that a large change in column polarity may result in a change in elution order. Failure to confirm the individual retention times of a series of solutes on a column of different polarity could result in misidentifications and serious errors in analysis.

The chemist's problem is to predict retention behavior for solutes while lacking a good system for specifying polarity. We saw in Chapter 3 that the adjusted retention volume is directly proportional to the distribution constant K_c, so it could serve as a measure of polarity, but distribution constants are not generally known. The best we can do within the context of this brief text is to discuss some of the basic principles of polarity based on our knowledge of intermolecular forces.

Polarity and Intermolecular Forces

Defining the polarity of a stationary phase is complicated and not easily quantified. Polarity is determined by intermolecular forces which are complex and difficult to predict in chromatographic systems. The polarity of a pure liquid can be specified by its dipole moment. Other physical properties, such as boiling point and vapor pressure, reflect the extent of intermolecular forces. A large dipole moment and a high boiling point would reflect high polarity and strong intermolecular forces. However, these parameters relate to pure liquids, and in GLC, we are interested in intermolecular forces between two different molecules—a solute in the vapor state and a liquid stationary phase. Such a system is complicated and it is impossible at this time to produce a single numerical scale that can be used to represent all possible interactions.

Classically, intermolecular forces have been classified as *van der Waals forces* (listed in Table 4.2) and *hydrogen bonds.* Of the van der Waals forces, dispersion is present between all organic compounds, even nonpolar ones. Consequently, dispersion is not of much interest except when nonpolar hydrocarbons are the solutes. Induction and orientation forces give selectivity to chromatographic systems, and they cause the *polarity* we have been discussing. However, attempts by chromatographers to refine these generalizations of polarity into more useful parameters have not been of much practical value.

Hydrogen bonding is better understood and is evidenced only if one of the molecules has a hydrogen atom bonded to an electronegative atom like nitrogen or oxygen. Examples are alcohols and amines which can both donate and receive a hydrogen atom to form a hydrogen bond. Other molecules such as ethers, aldehydes, ketones, and esters can only accept protons—they have none to donate. Hence they can form hydrogen bonds only with donors such as alcohols and amines. Hydrogen bonds are rela-

TABLE 4.2 Classification of van der Waals Forces

Name	Interaction	Investigator
Dispersion	Induced dipole–induced dipole	London (1930)
Induction	Dipole–induced dipole	Debye (1920)
Orientation	Dipole–dipole	Keesom (1912)

tively strong forces and they are very important in chromatography; participating molecules are usually classified as hydrogen bond donors and/or hydrogen bond acceptors.

The strength of hydrogen bonding can also cause unwanted interactions. Solutes capable of hydrogen bonding can become attached to the walls of injection ports, solid supports, and column tubing. Often these adsorptions result in slow desorptions giving rise to nonsymmetrical peaks called *tailing peaks*. This undesirable asymmetry in peak shape can often be eliminated by derivatizing the surface hydroxyl groups on the walls and on solid support surfaces. Silanization of solid supports is discussed in Chapter 5.

The combined effect of all intermolecular forces cannot be treated theoretically to produce a "polarity" value for a given molecule. Rather, empirical measurements, and indices calculated from empirical measurements, have been devised to represent molecular polarity.

Separation Factor

The separation factor, α, is a parameter measuring relative distribution constants; its value can be determined from a chromatogram. For two adjacent peaks, the separation factor is the ratio of their relative adjusted retention volumes

$$\alpha = \frac{(V_R')_2}{(V_R')_1} = \frac{k_2}{k_1} = \frac{(K_c)_2}{(K_c)_1} \tag{1}$$

defined so that $(V_R')_2$ is the second eluting peak. As noted in equation 1, the separation factor is also equal to the ratio of retention factors or the ratio of distribution constants for the two peaks. As such, it represents the *relative* interaction between each of the solutes and the stationary phase and can be used to express the *relative* intermolecular forces and the magnitude of their similarity or difference. In practice, it tells us how difficult it is to separate these two solutes—the larger the value of α, the easier the separation. If $\alpha = 1.00$, there is no differential solubility and no separation is possible. To summarize: K_c and k are constants that indicate the extent of intermolecular forces between a solute and a stationary phase, while α expresses the *differential* solubility for two solutes on a given stationary phase.

The relationship between α and resolution is given by equation 2.

$$R_s = \left(\frac{\alpha - 1}{\alpha}\right) \left(\frac{k}{k + 1}\right) \left(\frac{\sqrt{N}}{4}\right) \tag{2}$$

Using this equation, and making reasonable assumptions, it can be calculated that a good, packed column is capable of resolving peaks with an α-

value of about 1.1; a capillary column, having a larger plate number, is required for resolution of solutes with smaller α-values, down to about 1.02.

Improving a separation can be accomplished by effecting changes in any of the three parameters, N, k or α. For packed columns, α is often the parameter with the greatest effect. Changing it is usually accomplished by changing the stationary phase and thereby changing the polarity. That is to say, if one has a poor separation on a packed column, she/he usually selects a different stationary phase.

This procedure will work on OT columns too, but OT columns have such high efficiencies (N-values) that column changing is less frequent. The third parameter, k, can be increased by lowering the column temperature, usually an effective strategy, especially with OT columns which operate at lower temperatures than comparable packed columns. However, increasing k above a value of 10 will not produce much gain in resolution and retention times will be longer.

It is interesting to compare the effects on resolution of changing these parameters. Equation 2 can be rearranged to calculate the plate number required to achieve a resolution of 1.0 with varying values of k and α:

$$N_{\text{req}} = 16 \left(\frac{\alpha}{\alpha - 1} \right)^2 \left(\frac{k_2 + 1}{k_2} \right)^2 \qquad (3)$$

Table 4.3 lists some typical values. It shows why capillary columns with their large plate numbers are required for difficult separations.

Kovats Retention Index

In order to establish a scale of polarity, one needs a reliable method for specifying and measuring the retention behavior of solutes. Parameters such as retention volume and retention factor would seem to be suitable, but they are subject to too many variables. Relative values are much better, and one such parameter originally defined by Kovats [2] has been well received. It uses a homologous series of n-paraffins as standards against

TABLE 4.3   Plate Number Required (N_{req}) for a Resolution of 1.0.*

k	Separation Factor					
	1.01	1.05	1.10	1.50	2.00	5.00
0.10	19,749,136	853,776	234,256	17,424	7,744	3,025
0.50	1,468,944	63,504	17,424	1,296	576	225
1.00	652,864	28,224	7,744	576	256	100
2.00	367,236	15,876	4,356	324	144	56
5.00	235,031	10,161	2,788	207	92	36

* Calculated from Equation 3.

which adjusted retention volumes are measured for solutes of interest. His choice of *n*-paraffins was based not only on their relative availability but also on their very low polarity and their freedom from hydrogen bonding.

The Kovats retention index, *I*, assigns a value of 100 times the number of carbons to each of the *n*-paraffins. Thus, hexane has a value of 600 and heptane 700 on all liquid phases. When a homologous series of hydrocarbons is chromatographed, the intermolecular forces are relatively constant and the separation is controlled primarily by differences in vapor pressure (as reflected in boiling points). The chromatogram which is produced shows a logarithmic relationship between carbon numbers and adjusted retention times, reflecting the trend in boiling points among the members of the homologous series. A linear relationship is exhibited when the *log* of the adjusted retention time (or volume) is plotted versus the Kovats index as shown in Figure 4.3.

To find the Kovats index for a given solute on a given stationary phase, a few members of the paraffin homologous series are chromatographed and plotted. Then the solute is run under the same conditions and its Index value is determined from the graph. It is best if the paraffins chosen bracket the retention volume of the analyte. If the flow rate is kept constant during the gathering of these data, then adjusted *retention times* can be plotted. Alternatively, the index can be calculated from equation 4,

$$I = \left[\frac{\log (V_N)_u - \log (V_N)_x}{\log (V_N)_{x+1} - \log (V_N)_x} \right] + 100x \tag{4}$$

where the subscript u stands for the unknown analyte and x and (x + 1) stand for the number of carbons in the paraffins eluted just before and just after the analyte, respectively.

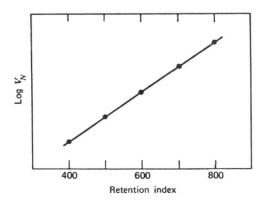

Fig. 4.3. Retention index (Kovats) plot. From Miller, J. M., *Chromatography: Concepts and Contrasts,* John Wiley & Sons, Inc., New York, 1987, p. 79. Reproduced courtesy of John Wiley & Sons, Inc.

The Kovats index has become a popular method for reporting GC data, replacing the absolute retention parameters. McReynolds [3] has published a reference book of self-consistent indices for 350 solutes on 77 stationary phases at two temperatures. From these data it can be seen that the Kovats index is not very temperature dependent and that adjacent members of *any* homologus series will have index values differing by about 100 units. Using this approximation, one can estimate the index for any chemical if the index for one member of its homologous series is known.

While the paraffins represent a set of universal standards for establishing an index, other homologous series have been used in particular industries where other series are commonly used [4]. For example, four index systems have been compared recently for characterizing nitrogenous acidic and neutral drugs [5]. The alkylhydantoins and alkylmethylhydantoins turned out to be the most feasible retention index standards for the compounds studied.

Rohrschneider–McReynolds Constants

Let us return to our discussion regarding the determination of the polarity of stationary phases by beginning with an example using Kovats retention indexes. From McReynolds [3] we find that toluene has a Kovats retention index of 773 on the nonpolar phase squalane and 860 on the more polar dioctylphthalate. The difference in these indexes, 87, provides a measure of the increased relative polarity of dioctylphthalate relative to squalane. The difference can be designated as ΔI.

Rohrschneider [6] proposed a list of five chemicals that could be used as test probes (like the solute toluene) to compare retention indexes on squalane (the universal nonpolar standard) and any other liquid phase. His choices are listed below (McReynolds' probes are also listed).

Probes Used By	
Rohrschneider	McReynolds
Benzene	Benzene
Ethanol	*n*-Butanol
2-Butanone (MEK)	2-Pentanone
Nitromethane	Nitropropane
Pyridine	Pyridine
	2-Methyl-2-pentanol
	Iodobutane
	2-Octyne
	1,4-Dioxane
	cis-Hydrindane

All five probes are run on squalane and on the stationary phase whose polarity is to be determined, and a set of five ΔI values are determined. Each serves to measure the extent of intermolecular interaction between the probe and the stationary phase, and together they provide a measure of the polarity of the stationary phase. More details about this procedure can be found in the paper by Supina and Rose [7].

In 1970 McReynolds [8] went one step further. He reasoned that ten probes would be better than five and that some of the original five should be replaced by higher homologs. It has turned out that ten probes and hence ten index values are too many. Most compilations of Rohrschneider–McReynolds values list only 5. Table 4.4 gives the ΔI values for 13 stationary phases.

Are McReynolds numbers of any use in specifying polarity? The arrangement in Table 4.4 is according to increasing value of the average of the five numbers and clearly shows that the polarity increases as one goes down the table. But how much? That is where the system falls short. Any one value can indicate a particularly strong interaction. For example, tricresylphosphate has an unusually high value for n-butanol, indicating that it interacts strongly with alcohols, probably by forming hydrogen bonds.

Are there any uses for the McReynolds System? Consider OV-202 and OV-210. They have identical values indicating that these two polymers are

TABLE 4.4 McReynolds Constants and Temperature Limits for Some Common Stationary Phases

Stationary Phase	Probes*					Temp. Limits	
	Benz	Alc	Ket	N-Pr	Pyrid	Lower	Upper
Squalane	0	0	0	0	0	20	125
Apolane 87®	21	10	3	12	25	20	260
OV-1®	16	55	44	65	42	100	375
OV-101®	17	57	45	67	43	20	375
Dexsil 300®	41	83	117	154	126	50	450
OV-17®	119	158	162	243	202	20	375
Tricresylphosphate	176	321	250	374	299	20	125
QF-1	144	233	355	463	305	0	250
OV-202® and OV-210®	146	238	358	468	310	0	275
OV-225®	228	369	338	492	386	20	300
Carbowax 20M®	322	536	368	572	510	60	225
DEGS	492	733	581	833	791	20	200
OV-275®	629	872	763	1106	849	20	275

* Benz = Benzene
 Alc = *n*-Butanol
 Ket = 2-Pentanone
 N-Pr = Nitropropane
 Pyrid = Pyridine
 ® = Registered Trademark

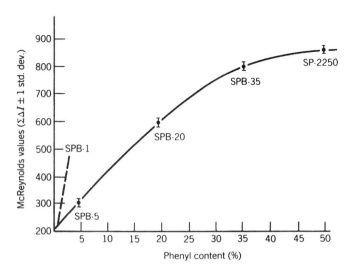

Fig. 4.4. Effect of number of phenyl groups on stationary phase polarity as measured by McReynolds values. Reprinted with permission of Supelco Inc., Bellefonte, PA, from the *Supelco Reporter,* Vol. IV, No. 3, May 1985. From Miller, J. M., *Chromatography: Concepts and Contrasts,* John Wiley & Sons, Inc., New York, 1987, p. 140. Reproduced courtesy of John Wiley & Sons, Inc.

identical except for differences in chain length and viscosity (which have little affect on polarity). This type of comparison was important in the early days of GC when new polymers were made to replace exhausted supplies of old polymers; for example OV-210 replaced QF-1. The McReynolds values provided proof of their equivalency.

Also, the sum of these five McReynolds values has been used to verify the increase in polarity of silicone polymers containing increasing percentages of phenyl groups. Figure 4.4 shows a plot for five silicone polymers on bonded fused silica WCOT columns (except for SP-2250, which is from packed column data). These examples show some utility for the method, but clearly we are still lacking a simple means for selecting a good stationary phase for a given separation.

Other Studies

Various groups of workers have attempted to refine or extend the empirical data of McReynolds by using a variety of theoretical approaches. Most have assumed that three or four types of intermolecular forces would be sufficient to characterize stationary phases: dispersion forces, dipolar interactions, and one or two types of hydrogen bonding. These efforts have not had much impact on the process of selecting stationary phases and will not be described further here. For further information, the works of Hartkopf [9], Hawkes [10], Snyder [11], Risby [12], and Carr [13] can be consulted.

Activity Coefficients

There is one other common way to express the interaction between a solute and a stationary phase and it arises from a consideration of the thermodynamics of solutions.

Raoult's Law expresses the relationship between the vapor pressure above a solution, p_A and the vapor pressure of a pure solute, p_A^0,

$$p_A = X_A p_A^0 \tag{5}$$

where X_A is the mole fraction of the solute A. Solutes being analyzed by GC often exhibit less than ideal behavior and follow Henry's Law in which a proportionality constant replaces the vapor pressure of pure solute. To allow for this non-ideality, Raoult's Law can be modified by introducing the concept of an activity coefficient, γ:

$$p_A = \gamma_A X_A p_A^0 \tag{6}$$

Thus the activity coefficient bears a relationship to the intermolecular forces between the solute and solvent. If it could be measured, it too would provide a measure of these forces.

Equation 7 shows the relationship between the activity coefficient and the distribution constant, K_c:

$$K_c = \frac{V_s d_s \mathcal{R} T}{\gamma\, p^0 (MW)_s} \tag{7}$$

$\mathcal{R}$ is the gas constant, T is the temperature, d_s is the density of the stationary phase, and $(MW)_s$ is the molecular weight of the stationary phase.

Consider two solutes, A and B, being chromatographed. The ratio of their distribution constants is equal to the ratio of their adjusted retention volumes as expressed in Equation 1, and substituting equation 7 into equation 1 yields:

$$\alpha = \frac{(K_c)_B}{(K_c)_A} = \frac{p_A^0\, \gamma_A}{p_B^0\, \gamma_B} \tag{8}$$

Since α expresses the extent of separation of A and B, equation 8 shows that this separation is dependent on two factors: the ratio of the vapor pressures (or boiling points), and the ratio of activity coefficients (or intermolecular forces between the solute and the stationary phase). It is for this reason that these two parameters were specified in Chapter 1 as the two important variables in setting up a GC system. It is the ratio of activity coefficients that gives GC its enhanced ability to achieve separations compared with distillation which is dependent only on vapor pressure ratios.

One classic example of the separation of two solutes with nearly the same boiling points is that of benzene (b.p. 80.1°C) and cyclohexane (b.p. 81.4°C). Even though they have very similar boiling points (and vapor pressures), they are easily separated by GC using a stationary liquid phase that has moderate polarity and interacts more strongly with the pi-cloud of benzene than it does with the less polar cyclohexane:

$$\alpha = \frac{p_{CY}^0 \, \gamma_{CY}}{p_{BZ}^0 \, \gamma_{BZ}} \cong \frac{\gamma_{CY}}{\gamma_{BZ}} \tag{9}$$

Activity coefficients can be calculated from GC data according to equation 10:

$$\gamma = \frac{1.7 \times 10^5}{V_g \, p^0 \, (MW)_s} \tag{10}$$

where V_g is the *specific* retention volume (the net retention volume at 0°C and per gram of stationary phase). When benzene and cyclohexane are chromatographed on dinonylphthalate at 325°K, their activity coefficients are found to be 0.52 and 0.82 respectively [14] and $\alpha = 1.6 = 0.82/0.52$. Benzene is retained more than cyclohexane by the polar dinonylphthalate because of its larger intermolecular interactions. While activity coefficients are not commonly determined for this purpose, it is clear that they are a valid means for expressing intermolecular interactions in GC.

LIQUID STATIONARY PHASES (GLC)

Squalane has already been discussed as the liquid phase considered to have the least polarity. It is a saturated hydrocarbon with the formula $C_{30}H_{62}$; its structure is shown in Figure 4.5. Its upper temperature limit is only 125°C, so a larger paraffin, Apolane 87, with the formula $C_{87}H_{176}$ has often been used as a substitute even though it is slightly more polar (see Table 4.4).

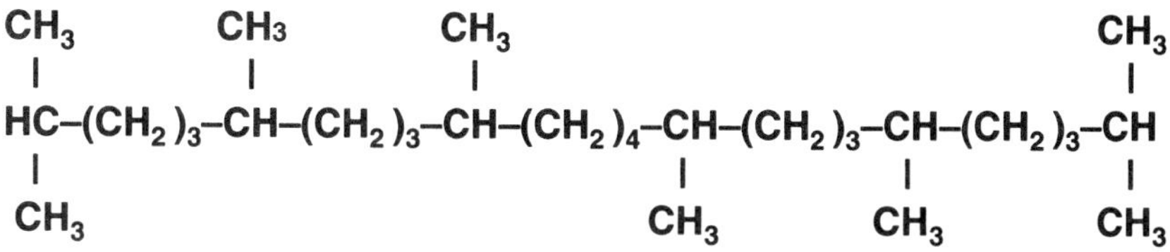

Fig. 4.5. Structure of squalane, a saturated, highly branched C_{30} hydrocarbon.

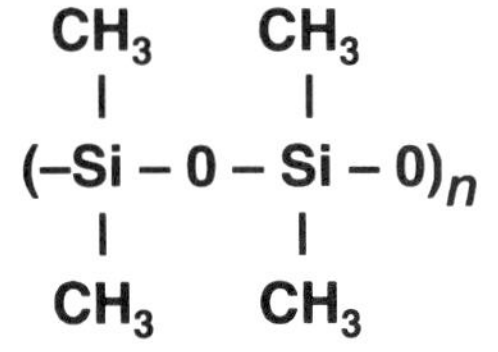

$$
\begin{array}{cc}
CH_3 & CH_3 \\
| & | \\
(-Si - 0 - Si - 0)_n \\
| & | \\
CH_3 & CH_3
\end{array}
$$

Fig. 4.6. Structure of OV-1® or DB-1®, a dimethylpolysiloxane.

Silicone Polymers

Silicone polymers have good temperature stability and modified silicone polymers now dominate the commonly used liquid phases. A range of polarities can be provided by changing the percentage of polar groups, for example phenyl and cyanopropyl groups. The least polar is a dimethylsilicone whose structure is shown in Figure 4.6; it is sold under the names OV-1 and OV-101 by Ohio Valley Specialty Chemical, the former being a gum and the latter a liquid. Both are included in a complete listing of silicone phases in Appendix VI.

From the McReynolds constants (Table 4.4), it can be seen that OV-1 and OV-101 have essentially the same polarity and are slightly more polar than Apolane 87. As the methyl groups are replaced by the more polar phenyl and cyanopropyl groups, the polarities increase as evidenced by the increasing McReynolds constants. Table 4.5 lists some alternative designations used for these polymers by other manufacturers.

Other Common Phases

Even this long listing of silicone polymers does not meet the needs of all chromatographers who seek liquids with higher polarity and/or higher operating temperatures. A series of polyethyleneglycol polymers has met some of the need for higher polarity since these materials can hydrogen bond. The structure of these polymers is given in Figure 4.7. The approximate molecular weight is given as a numerical value in the name. For example Carbowax 20M® has a molecular weight of 20,000; it is the highest

TABLE 4.5 Equivalent Silicone Polymer Liquid Phases*

Ohio Valley Number	Other Designations					
OV-1, 101	SP2100	SPB-1	DB-1	HP-1	SE-30	DC-200
OV-73	—	SPB-5	DB-5	HP-5	SE-52	SE-54
OV-17	SP-2250	SPB-50	DB-17	HP-17	—	—
OV-202, 210	SP-2401	—	DB-210	—	—	QF-1
OV-275	SP-2340	—	—	—	—	CP-Sil 88

* In order of increasing McReynolds values.

$$OH-(-CH_2-CH_2-O)_n-H$$

Fig. 4.7. Structure of Carbowax ® 20M, a polymeric polyethyleneglycol with an average molecular weight of 20,000.

molecular weight commercially available and can be used up to 225°C in packed columns and 280°C in some bonded capillary columns.

A series of carborane silicone polymers (see Fig. 4.8) has been designed especially for high temperature work. They were first synthesized in 1964 by Olin Chemical and are sold under the trade name Dexsil.® Dexsil 300® is the least polar, having all methyl groups in the chain. It can be used up to 400°C.

Recommended Stationary Phases

For practical reasons it is desirable to have only the minimum number of columns that will solve one's most frequent separation problems. Open tubular columns are so efficient that fewer of them are usually needed, but it is common to have two to four different phases and several different film thicknesses and lengths. More specific information is given in each of the respective chapters on packed and capillary columns.

Choosing a Stationary Liquid Phase

It is clear from the foregoing discussion that no convenient system has been found for guiding the selection process. Certainly one cannot rely on McReynolds constants alone. The simple maxim commonly used is the one with which we began this section—"like dissolves like." Which is to say, one chooses a nonpolar column for a nonpolar mixture and a polar column for a polar mixture.

An exception to this generalization occurs when one attempts to separate similar solutes such as isomers. For example, the xylene isomers are all more or less nonpolar and have similar boiling points. A nonpolar station-

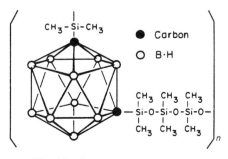

Fig. 4.8. Structure of Dexsil 300.®

ary phase will not be satisfactory for their separation because they do not vary much in either boiling point or in polarity. To accentuate the *small* differences in polarity requires a polar stationary phase like DB-WAX.® Figure 4.9 shows a good separation of this challenging separation.

Following the general polarity rule and using packed columns, a lightly loaded (5%), four-foot, 2 mm i.d., OV-101 column would be a good choice for a nonpolar sample and a similar Carbowax 20M® column for a polar sample. Between these two extremes, one of the silicone polymers of inter-mediate polarity (such as OV-17) could be used. Other special packings are discussed in Chapter 5.

For open tubular columns the choice of stationary phase is much less critical. A thin film (0.25 μm) 15-meter methyl silicone column (OV-101)

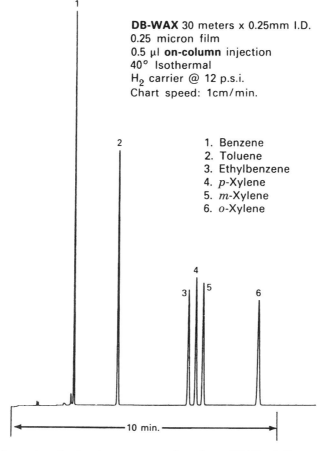

DB-WAX 30 meters x 0.25mm I.D.
0.25 micron film
0.5 μl **on-column** injection
40° Isothermal
H_2 carrier @ 12 p.s.i.
Chart speed: 1cm/min.

1. Benzene
2. Toluene
3. Ethylbenzene
4. *p*-Xylene
5. *m*-Xylene
6. *o*-Xylene

◄──────── 10 min. ────────►

Fig. 4.9. Separation of xylene isomers on a polar column, DB-Wax.® Courtesy of J & W Scientific, Inc. From Miller, J. M., *Chromatography: Concepts and Contrasts,* John Wiley & Sons, Inc., New York, 1987, p. 48. Reproduced courtesy of John Wiley & Sons, Inc.

TABLE 4.6 Some Common GC Adsorbents

	Commercial Trade Names
Silica gel	Davidson Grade 12, Chromasil®, Porasil®
Activated alumina	Alumina F-1, Unibeads-A®
Zeolite molecular sieves	MS 5A, MS 13X
Carbon molecular sieves	Carbopack®, Carbotrap®, Carbograph®, Graphpac®
Porous polymers	Porapak®, HayeSep®, Chromosorb® Century Series
Tenax polymer	Tenax TA®, Tenax GR®

would be good for general screening. A similar, but more polar silicone polymer (for example the cyano-derivatives, OV-225 or OV-275) would be better for more polar samples. The capillary column equivalent of Carbowax 20M® (such as DB-WAX®) is a logical choice also, even though these columns are easily oxidized and have relatively short useful lifetimes.

A final consideration in choosing a liquid phase is its temperature limitations. At the upper end, a temperature is reached where the vapor pressure of the liquid is too high and it bleeds off the column giving a high background detector signal. At such high temperatures the column lifetime is short and the chromatography is poor due to the bleed. The tables of common liquid phases in this chapter have included these upper temperature limits for the phases when used in packed columns. The limits for open tubular columns are similar—usually a bit higher if bonded to the column. The lower temperature limit is usually the freezing point or glass transition temperature of the polymer. The classic example is Carbowax® 20M which is a solid at room temperature and melts around 60°C, its lower temperature limit.

SOLID STATIONARY PHASES (GSC)

Solids used in GSC are traditionally run in packed columns, the subject of the next chapter. However, a list of common solids is given in Table 4.6. As with the solid supports used in GLC, these solids should have small particle sizes and be uniform—for example 80/100 mesh range.

Some of these solids have been coated on the inside walls of capillary columns and are called support coated open tubular or SCOT columns. More information about them is included in Chapter 6.

REFERENCES

1. Miller, J. M., *J. Chem. Educ.,* **41,** 413 (1964).

2. Kovats, E. S., *Helv. Chim. Acta,* **41,** 1915 (1958).

3. McReynolds, W. O., *Gas Chromatographic Retention Data,* Preston Technical Abstracts, Evanston, IL, 1966.

4. Blomberg, L. G., *Advances in Chromatography,* Vol. 26, J. C. Giddings, Ed., Dekker, NY, 1987. Chpt. 6.

5. Rasanen, I., Ojanpera, I., Vuori, E., and Hase, T. A., *J. Chromatogr. A,* **738,** 233 (1996).

6. Rohrschneider, L., *J. Chromatogr,* **22,** 6 (1966).

7. Supina, W. R., and Rose, L. P., *J. Chromatogr. Sci.,* **8,** 214 (1970).

8. McReynolds, W. O., *J. Chromatogr. Sci.,* **8,** 685 (1970).

9. Hartkopf, A., *J. Chromatogr. Sci.,* **12,** 113 (1974); Hartkopf, A., Grunfeld, S., and Delumyea, R., *idem,* 119.

10. Hawkes, S., et al., *J. Chromatogr. Sci.,* **13,** 115 (1975); Burns, W., and Hawkes, S., *J. Chromatogr. Sci.* **15,** 185 (1977); Chong E., deBriceno, B., Miller, G., and Hawkes, S., *Chromatographia,* **20,** 293 (1985); Burns, W., and Hawkes, S. J., *J. Chromatogr. Sci.* **15,** 185 (1977).

11. Snyder, L. R., *J. Chromatogr. Sci.,* **16,** 223 (1978); Karger, B. L., Snyder, L. R., and Eon, C., *Anal. Chem.,* **50,** 2126 (1978).

12. Figgins, C. E., Reinbold, B. L., and Risby, T. H., *J. Chromatogr. Sci.* **15,** 208 (1977).

13. Li, J., Zhang, Y., and Carr, P., *Anal. Chem,* **64,** 210 (1992).

14. Kenworthy, S., Miller, J., and Martire, D. E., *J. Chem. Educ.,* **40,** 541 (1963).

5. Packed Columns and Inlets

All initial work in gas chromatography was performed on packed columns and the first commercial instruments accepted only packed columns. Later, when open tubular capillary columns were *invented,* only one manufacturer (Perkin-Elmer) produced them, so most chromatographers continued to use packed columns. As a result, much of the early literature reports only packed column separations. Today, however, it is estimated that over 80% of all analyses are made on capillary columns.

The two types of columns are sufficiently different that each will be discussed in a separate chapter. The instrumental requirements are also somewhat different, so the respective inlet systems for each type are included in each of the chapters.

Packed columns are typically made of stainless steel and have outside diameters of 1/4 or 1/8 inch and lengths of 2 to 10 feet. For applications requiring greater inertness, alternative materials have been used including glass, nickel, fluorocarbon polymers (Teflon®), and steel that is lined with glass or Teflon®. Copper and aluminum are conveniently soft for easy bending, but are not recommended due to their reactivity.

SOLID SUPPORTS

For packed columns, the stationary liquid phase is coated on a *solid support* which is chosen for its high surface area and inertness. Many materials

TABLE 5.1   Representative Solid Supports*

Name	Surface Area (m²/g)	Packed Density (g/cc)	Pore Size (μm)	Maximum % Liquid Phase
Diatomaceous Earth Type				
Chromosorb P®	4.0	0.47	0.4–2	30
Chromosorb W®	1.0	0.24	8–9	15
Chromosorb G®	0.5	0.58	NA	5
Chromosorb® 750	0.7	0.40	NA	7
Fluorocarbon Polymer				
Chromosorb® T	7.5	0.42	NA	10

* Manufactured by and exclusive trade marks of the Celite Corp.
NA = Not available

have been used, but those made from diatomaceous earth (Chromosorb®) have been found to be best. The properties of the major types are listed in Table 5.1.

The surfaces of the diatomaceous earth supports are often too active for polar GC samples. They contain free hydroxyl-groups that can form undesirable hydrogen bonds to solute molecules and cause tailing peaks. Even the most inert material (white Chromosorb W®) needs to be acid washed (designated AW) and silanized to make it still more inert [1]. Some typical silanizing reagents are dimethyldichlorosilane (DMDCS) and hexamethyldisilazane (HMDS). The deactivated white supports are known by names such as Supelcoport®, Chromosorb W-HP®, Gas Chrom Q II®, and Anachrom Q®. One disadvantage of deactivation is that these supports become hydrophobic, and coating them with a polar stationary liquid can be difficult.

As noted in Chapter 3, narrow ranges of small particles produce more efficient columns. Particle size is usually given according to mesh range, determined by the pore sizes of the sieves used for screening (see Table 5.2). Common choices for GC are 80/100 or 100/120 mesh.

The amount of liquid phase coated on the solid support varies with the support and can range from 1 to 25%. Table 5.3 shows that 15% liquid phase on Chromosorb P® is a loading equivalent to nearly twice that amount (25.7%) on Chromosorb W® due to their differences in density and surface

TABLE 5.2   Mesh and Particle Sizes

Mesh Range*	Particle Diameter (μm)		Range (μm)
	From	To	
80/100	177	149	28
100/120	149	125	24

* For a definition of mesh see Chapter 3

TABLE 5.3   Equivalent Stationary Phase Loadings (in weight percent) for Three Solid Supports.

Chromosorb P®	Chromosorb W®	Chromosorb G®
5.0	9.3	4.1
10.0	17.9	8.3
15.0	25.7	12.5
20.0	32.8	16.8
25.0	39.5	21.3
30.0	45.6	25.8

Taken from: Durbin, D. E., *Anal. Chem.* **45,** 818 (1973). Reprinted with permission from *Analytical Chemistry,* Copyright 1973, American Chemical Society.

area. Chromosorb G® can only hold small amounts of liquid (typically 3–5%).

Low loadings are better for high efficiency and high boiling compounds, and the high loadings are better for large samples or volatile solutes—gases for example. A solution of the stationary phase is made in a volatile solvent, mixed with the solid support, and evaporated to remove the solvent. The final material, even those with 25% liquid stationary phase (on Chromosorb P®) will appear *dry* and will pack easily into the column.

LIQUID STATIONARY PHASES

Virtually every nonvolatile liquid found in a common chemical laboratory has been tested as a possible stationary phase. As a consequence, there are too many liquid phases listed in commercial suppliers' catalogs (typically about 200 of them). The problem is to restrict the long list of phases to a few which will solve most analytical problems. To this end several workers have published their lists of preferred phases. A few of these choices are listed in Table 5.4. In general, they include a nonpolar column like the

TABLE 5.4   Recommended Liquid Stationary Phases

Hawkes[a]	Yancey[b]	NcNair[c]
OV-101	OV-101	OV-1
OV-17	OV-17	OV-17
Carbowax® ≥ 4000	Carbowax 20M®	Carbowax 20M®
OV-210	OV-202	OV-210
DEGS	OV-225	OV-275
Silar 10C		

[a] Hawkes, S., et al., *J. Chromatogr. Sci.* **13,** 115 (1975).

[b] Yancey, J. A., *J. Chromatogr. Sci.* **24,** 117 (1986).

[c] McNair, H. M., *ACS Training Manual,* American Chemical Society, 1994.

methyl silicones, several of intermediate polarity, a highly polar silicone like OV-275, and a polyglycol like Carbowax®.

A secondary consideration is the amount of stationary phase needed to coat the solid support. Table 5.1 listed the upper limits for some supports. The lower limit is usually the minimum amount that will give complete coverage of the support surface, an amount that is dependent on the surface area (also listed in Table 5.1). However, uniform coatings are difficult to attain especially for polar liquids, and the minimum percentage is usually determined by trial and error.

A third consideration is column length, but it is not critical if the instrument is capable of programmed temperature (see Chapter 9). Column lengths are usually short (1 to 3 meters) for convenience in both packing and handling.

Choosing the best column (liquid phase) for a given sample was discussed in Chapter 4, but a useful reference providing nearly 200 examples of actual separations on packed columns has been made available recently by Supelco [2]. Other suppliers also provide application information.

SOLID STATIONARY PHASES (GSC)

Common adsorbent solids like silica gel and alumina are used in GSC, but most of the solids used as stationary phases have been developed for specific applications in GSC. Table 4.6 listed some of them and this chapter will illustrate some of their common uses.

A typical separation of fixed gases on silica gel is shown in Figure 5.1. Although the peak shapes and plate number are rather good in this example, many of the solids used in GSC produce poor shapes (usually tailing) and disappointing efficiencies. Note that air is not separated into oxygen and nitrogen on silica gel.

It is easy to separate oxygen and nitrogen using solids known as molecular sieves, naturally occurring Zeolites and synthetic materials like alkali metal aluminoscilicates. The classic separation on a synthetic molecular sieve is shown in Figure 5.2. These sieves are named in accordance with their approximate effective pore sizes, e.g., 5A has 5 Å pores and 13X, 9 Å pores. The separation of oxygen and nitrogen is about the same on either sieve, but CO takes twice as long to elute from the 5 Å molecular sieve.

Carbosieves® are typical of solids that have been made for GC, in this case by pyrolysis of a polymeric precursor that yields pure carbon containing small pores and serving as a molecular sieve. The Carbosieves® will separate oxygen and nitrogen and can be substituted for the molecular sieves just described. They also find use for the separation of low molecular weight hydrocarbons and formaldehyde, methanol and water. Other trade names are Ambersorb® and Carboxen®.

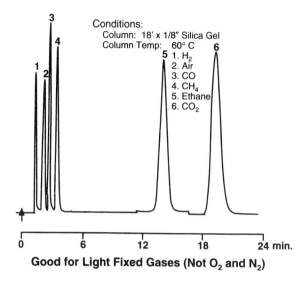

Conditions:
Column: 18' x 1/8" Silica Gel
Column Temp: 60° C
1. H$_2$
2. Air
3. CO
4. CH$_4$
5. Ethane
6. CO$_2$

Good for Light Fixed Gases (Not O$_2$ and N$_2$)

Fig. 5.1. Separation of light fixed gases on silica gel.

Another class of carbon adsorbents are graphitized carbon blacks which are nonporous and nonspecific and separate organic molecules according to geometric structure and polarity. Often they are also lightly coated with a liquid phase to enhance their performance and minimize tailing. Figure 5.3 shows a typical separation of a solvent mixture. One common trade name for these materials is Carbopack®.

In 1996 Hollis [3] prepared and patented a porous polymer that has been marketed under the trade name Porapak®. It provided a good solution to the problem of separating and analyzing water in polar solvents. Because of its strong tendency to hydrogen-bond, water usually tails badly on most stationary phases, but Porapak® solves that problem as shown in Figure 5.4.

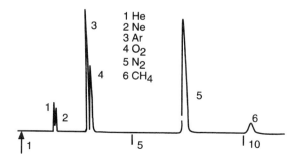

1 He
2 Ne
3 Ar
4 O$_2$
5 N$_2$
6 CH$_4$

Fig. 5.2. Separation of oxygen and nitrogen on molecular sieve column; 25-m PLOT.

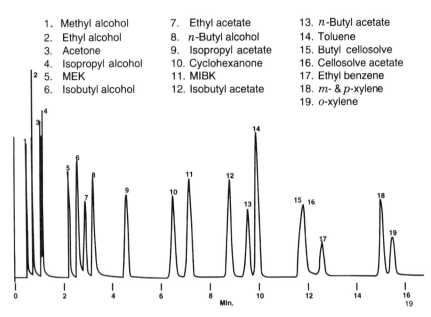

1. Methyl alcohol	7. Ethyl acetate	13. *n*-Butyl acetate
2. Ethyl alcohol	8. *n*-Butyl alcohol	14. Toluene
3. Acetone	9. Isopropyl acetate	15. Butyl cellosolve
4. Isopropyl alcohol	10. Cyclohexanone	16. Cellosolve acetate
5. MEK	11. MIBK	17. Ethyl benzene
6. Isobutyl alcohol	12. Isobutyl acetate	18. *m*- & *p*-xylene
		19. *o*-xylene

Fig. 5.3. Solvent separation on Carbopack®. Conditions: column, 6 ft. × 1/8″ O.D. SS, Carbopack C coated with 0.1% SP-1000; temperature program, 100°C to 225°C @ 8°C/min; flow, 20 mL/min nitrogen; detector, FID.

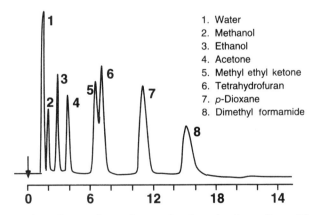

1. Water
2. Methanol
3. Ethanol
4. Acetone
5. Methyl ethyl ketone
6. Tetrahydrofuran
7. *p*-Dioxane
8. Dimethyl formamide

Fig. 5.4. Separation of water in a mixture of polar solvents on Porapak® Q. Column 6 ft. × 1/4″ OD, 150/200 mesh Porapak® Q, at 220°C; flow rate 37 mL/min He; TCD.

Originally there were five different polymers, designated P through T in increasing polarity; now there are eight versions. Water elutes very quickly on Porapak P and Q making them ideal for those applications where water would otherwise interfere with compounds of interest. Porapak Q® can also be used to separate oxygen and nitrogen at −78°C. A competitive series of polymers is sold under the trade name Chromosorb® Century Series. For further examples of applications consult the literature available from chromatographic supply houses (see Appendix VIII).

GAS ANALYSIS

The analysis of gases is one of the major applications of packed column GSC. The characteristics of packed columns that make them ideal for gas analysis are:

* adsorbents provide high surface areas for maximum interaction with gases that may be difficult to retain on liquid stationary phases;

* large samples can be accommodated, providing lower absolute detection limits;

* some packed column GCs can be configured to run below ambient temperature which will also increase the retention of the gaseous solutes;

* unique combinations of multiple columns and/or valving that make it possible to optimize a particular sample. Figure 5.5 shows one such application for shale oil gases.

Gas sampling valves are also common on these instruments. A common configuration is a 6-port valve shown in Figure 5.6. It is operated in one of two positions: one for filling the sample loop and one for "injecting" the sample. There is essentially no dead volume with gas sample valves and the repeatability is very good.

Valves can be contained in separate ovens to assure reproducible quantitative sampling. The sample pressure in these valves is important for accurate quantitation. However, if the sample loop is at ambient pressure and the column inlet is at elevated pressure as required for the analysis, a rather large baseline shift is often observed as shown in Figure 5.7. Consequently, sample loops are often filled at the higher pressures to eliminate this problem. Alternatively, the column can be operated at constant pressure instead of the more common constant flow.

Valves can also be used for column switching to achieve unique configurations for specific separations. A review by Willis [4] contains many different valving arrangements. Backflushing can also be achieved with proper valving and it is a commonly applied technique in gas analysis.

Gas chromatographs used for gas analysis usually are fitted with thermal conductivity detectors (TCD), which are universal, stable, and moderately sensitive and are usually run with helium carrier gas. Since most thermal conductivity detectors are differential and have two active elements, both

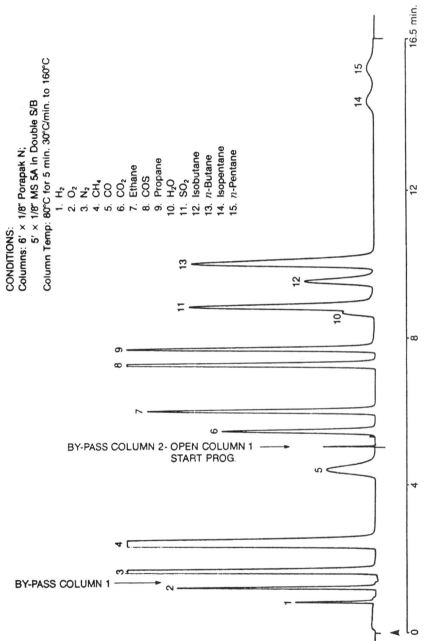

CONDITIONS:
Columns: 6' × 1/8" Porapak N;
5' × 1/8" MS 5A in Double S/B
Column Temp: 80°C for 5 min. 30°C/min. to 160°C

1. H_2
2. O_2
3. N_2
4. CH_4
5. CO
6. CO_2
7. Ethane
8. COS
9. Propane
10. H_2O
11. SO_2
12. Isobutane
13. n-Butane
14. Isopentane
15. n-Pentane

BY-PASS COLUMN 2 - OPEN COLUMN 1 ⟶
START PROG.

BY-PASS COLUMN 1 ⟶

Fig. 5.5. Multicolumn separation of oil shale gas. Reprinted from reference [6] courtesy of Varian Analytical Instruments.

80

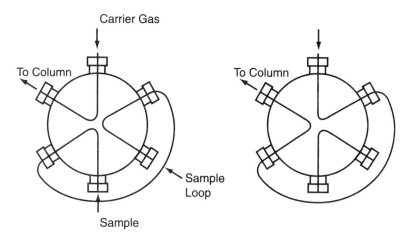

(*a*) Sample Load Position   (*b*) Sample Inject Position

Fig. 5.6. Typical 6-port sample valve: (*a*) load position; (*b*) inject position.

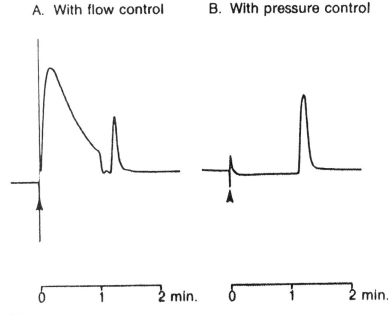

Fig. 5.7. Effect of injection on baseline: (*a*) with flow control; (*b*) with pressure control. Reprinted from reference [6] courtesy of Varian Analytical Instruments.

elements can be used for specialized column arrangements including dual column operation.

One type of sample that cannot be analyzed with a TCD and helium carrier gas is hydrogen in a gas mixture. Hydrogen's thermal conductivity is so close to helium's that the peak shapes are often irregular—usually with a W-shape—and thus quantitative results are not possible [5]. The thermal conductivity of binary mixtures of helium and hydrogen is not a simple linear function. More discussion and some possible solutions to this problem can be found in Thompson's monograph [6].

When a more sensitive detector is needed, the TCD is inadequate and the FID is often not satisfactory because it is not universal. In this situation, such as might occur in environmental gas analysis, one of the other ionization detectors is preferable. There are commercially available ionization detectors that have met this need [7].

INLETS FOR LIQUID SAMPLES AND SOLUTIONS

Sample introduction for liquids on packed columns is often accomplished with a microsyringe through a self-sealing silicone septum as shown in Figure 5.8. In this figure, the column is lined up colinearly with the syringe needle providing for either of two possible injection modes: on-column or flash vaporization.

For on-column operation, the column is positioned as shown in Figure 5.8 with the column packing beginning at a position just reached by the needle. When the syringe is pushed as far as it will go into the port, its contents will be delivered into the first part of the column packing—ideally on a small glass wool plug used to hold the packing in the column. There the analytes will be sorbed onto the column or evaporated depending on

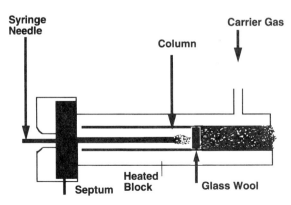

Fig. 5.8. Simplified injection port for on-column injection.

their relative distribution constants. For most samples, the majority of the sample will go into the stationary phase (see the calculations in Chapter 3); hence the name *on column*. When purchasing a commercial column for on-column injection, it is necessary to specify the length of column that should be left empty in accordance with the geometric requirements just discussed. An alternative is to use a pre-column liner which can be replaced or cleaned when it gets dirty.

In the second configuration, the column is placed so that its front end (and its packing) barely extends into the injection port and cannot be reached by the syringe needle. Efficient sampling for this configuration requires that the sample evaporate quickly (flash vaporization) when injected into the port. This operation is facilitated by heating the injection port to a temperature well above the boiling point of the sample to insure rapid volatilization. One possible disadvantage of this method is that the sample will probably come into contact with the hot walls of the port and may undergo thermal decomposition. For this reason, inert glass liners are often inserted into the injection port.

Packed column GCs are almost always operated at constant flow of carrier gas. A valve for this purpose is essential for programmed temperature work. Constant flow operation is preferred for thermal conductivity detectors as explained in the detector chapter (Chapter 7).

SPECIAL COLUMNS

Special columns include those for specific analyses which cannot be easily accomplished with the common packings and those with unusual dimensions like the so-called microbore columns.

Special Applications

There are a number of stationary phases that have been designed to provide special selectivity for difficult analyses. Some are listed in Table 5.5.

It has been found that mixing several different liquid phases in one column will produce a selectivity directly proportional to the sum of the parts mixed together [8–12]. Usually it does not matter if the phases are kept separate in the column or are mixed together. A few useful ones are listed in Table 5.5. Some are commercially available, for example, those used in EPA methods for wastewater analysis. Since this flexibility cannot easily be attained with capillary columns, mixed packings represent one of the unique advantages of packed columns.

As a general rule, highly acidic or highly basic samples are difficult to chromatograph because of their high reactivity and strong hydrogen-bonding. To counteract these effects it has become common to add a small amount (1–2%) of modifier to the liquid phase to cover up the most active

TABLE 5.5 Stationary Phase Packings for Special Applications

Column Packing	Application
A. Mixed Liquid Phases	
1.5% OV-17® + 1.95% OV-210®	Pesticides
2% OV-17® + 1% OV-210®	Amino acids
20% SP-2401® + 0.1% Carbowax 1500®	Solvents
1.5% OV-17® + 1.95% QF-1®	Phthalates, EPA method
B. Mixed Liquids (tail reducers) for Deactivation	
10% Apiezon L® + 2% KOH	Amines
12% Polyphenylether + 1.5% H$_3$PO$_4$ (on and in PTFE)	Sulfur gases
C. Other	
5% SP-1200® + 5% Bentone 34® clay	Xylene isomers
10% Petrocol A, B, or C	Simulated distillation
0.19% Picric Acid on Carbopack C®	Light hydrocarbons, unsaturated
5% Fluorcol on Carbopack B®	Freons

sites. For example, sodium or potassium hydroxide is used to deactivate the packing used for basic compounds like amines and phosphoric acid for acidic compounds like free acids and phenols.

Other special columns are listed in Table 5.5. Most are available commercially in packed columns. Chiral packings are another important type; they are more often used in capillary columns and they are discussed in Chapter 10.

Microbore Columns

Column performance improves as the column diameter decreases, but very small diameters represent a special case because of the packing difficulties and the high pressure drops that result. Packed columns with inside diameters of 750 micrometers are commercially available for a few phases. They are used when a compromise between normal packed columns and normal open tubular columns is needed. Some examples are: (a) to achieve both high efficiency and high sample capacity; (b) for highly volatile samples; (c) for greater speed than is possible with normal packed columns; or (d) to obtain the selectivity advantages of a mixed packing.

UPGRADING FOR CAPILLARY COLUMNS

While packed columns are preferred for some analyses, capillary columns are more efficient and are preferred for general use. Older, packed-column gas chromatographs can be retrofitted in the field to accept capillary columns, thereby upgrading them at minimal cost.

The major changes that are required are the installation of a capillary injector and the addition of make-up gas for the detector. Kits for this purpose are available from lab supply houses, and McMurtrey and Knight [13] have described the construction of a homemade one. The easiest conversion is from packed columns to wide-bore columns [14]; Jennings [15] has discussed the procedure in detail. The conversion is rather simple; as a minimum, all one needs are some fittings and tubing. These columns are useable with thermal conductivity detectors [16] as well as flame ionization detectors.

REFERENCES

1. Ottenstein, D. M., *J. Chromatogr, Sci.*, **11**, 136 (1973).

2. Anonymous, *Packed Column Application Guide, Bulletin 890*, Supelco, Bellefonte, PA, 1995.

3. Hollis, O. L., *Anal. Chem.*, **38**, 309 (1966).

4. Willis, D. E., *Advances in Chromatography*, Vol. 28, J. C. Giddings, Ed., Dekker, NY, 1989, Chpt 2.

5. Miller, J. M., and Lawson, A. E., Jr., *Anal. Chem.* **37**, 1348 (1965).

6. Thompson, B., *Fundamentals of Gas Analysis by Gas Chromatography*, Varian Associates, Palo Alto, CA, 1977.

7. Madabushi, J., Cai, H., Stearns, S., and Wentworth, W., *Am. Lab.* **27**, [15], 21 (1995).

8. Laub, R. J., and Purnell, J. H., *Anal. Chem.* **48**, 799 (1976).

9. Chien, C. F., Laub, R. J., and Kopecni, M. M., *Anal. Chem.* **52**, 1402 and 1407 (1980).

10. Lynch, D. F., Palocsya, and Leary, J. J., *J. Chromatogr. Sci.*, **13**, 533 (1975).

11. Pecsok, R. L., and Apffel, A., *Anal. Chem.* **51**, 594 (1979).

12. Price, G. J., *Advances in Chromatography*, Vol. 28, J. C. Giddings, Ed., Dekker, NY, 1989, Chpt. 3.

13. McMurtrey, K. D., and Knight, T. J., *Anal. Chem.*, **55**, 974 (1983).

14. Duffy, M. L., *Am. Lab*, **17**, [10], 94 (1985).

15. Jennings, W., *Analytical Gas Chromatography*, Academic Press, San Diego, 1987, Chapter 17.

16. Lochmuller, C. H., Gordon, B. M., Lawson, A. E., and Mathieu, R. J., *J. Chromatogr. Sci.* **16**, 523 (1978).

6. Capillary Columns and Inlets

Capillary columns were introduced in 1959, but were not used widely until about 1980, after which they grew steadily in popularity. Today, it is estimated that over 80% of all applications are run on capillary columns. Capillary columns are simply columns that are open tubes. That is, they are not filled with packing material. Instead, a thin film of liquid phase coats the inside wall. As discussed earlier, such columns are properly called "open tubular (OT) columns." Since the tube is open, its resistance to flow is very low, and long lengths, up to 100 meters are possible.

These long lengths make possible very efficient separations of complex sample mixtures. Figure 6.1 is a typical chromatogram of a standard text mixture on a 30-m fused silica OT. Notice the sharp symmetrical peaks obtained for polar, acidic, and basic compounds.

TYPES OF OT COLUMNS

The original capillary column, invented and patented by Dr. Marcel Golay [1], consisted of a tube with a thin film of liquid phase coated on the inside surface. This is properly called a wall-coated open tubular column or WCOT, shown in Figure 6.2. The tube can be made of fused silica, glass, or stainless steel. Almost all commercial capillary columns are now made of fused silica.

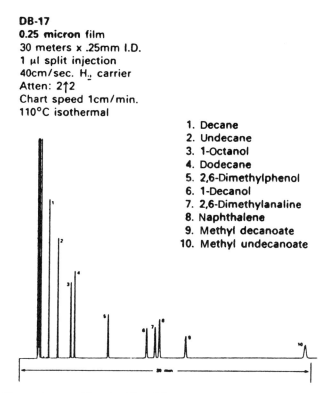

DB-17
0.25 micron film
30 meters x .25mm I.D.
1 µl split injection
40cm/sec. H₂ carrier
Atten: 2↑2
Chart speed 1cm/min.
110°C isothermal

1. Decane
2. Undecane
3. 1-Octanol
4. Dodecane
5. 2,6-Dimethylphenol
6. 1-Decanol
7. 2,6-Dimethylanaline
8. Naphthalene
9. Methyl decanoate
10. Methyl undecanoate

Fig. 6.1. Chromatogram of Standard Test Mixture. Conditions: 30 m × 0.25 mm ID DB-17 with 0.25 µm film at 110°C. Courtesy of J & W Scientific, Inc. From Miller, J. M., *Chromatography: Concepts and Contrasts,* John Wiley & Sons, Inc., New York, 1987, p. 118. Reproduced courtesy of John Wiley & Sons, Inc.

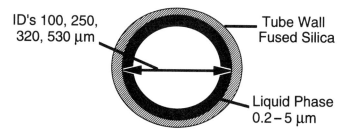

ID's 100, 250, 320, 530 µm

Tube Wall
Fused Silica

Liquid Phase
0.2 – 5 µm

Fig. 6.2. Wall coated open tubular (WCOT) column.

Wall-coated capillary columns provide the highest resolution of all gas chromatographic columns. Tubing internal diameters of 0.10, 0.20, 0.25, 0.32, and 0.53 millimeters are commercially available. Typical lengths vary from 10 to 50 meters, although 100 meter columns have been used occasionally and are commercially available. Long column lengths, however, do require long analysis times. Film coating thickness varies from 0.1 to 5.0 micrometers. Thin films provide high resolution and fast analysis, but they have limited sample capacity. Thicker films have higher sample capacity, but show lower resolution and are typically used for only very volatile compounds.

The other types of capillary columns are shown in Figure 6.3, the SCOT or support-coated open tubular column, on the left, and the PLOT or porous layer open tubular column, on the right. SCOT columns contain an adsorbed layer of very small solid support (such as Celite®) coated with a liquid phase. SCOT columns can hold more liquid phase, and have a higher sample capacity than the thin films common to the early WCOT columns. However, with the introduction of cross-linking techniques, stable thick films are possible for WCOT columns, and the need for SCOT columns has disappeared. A few SCOT columns are still commercially available but only in stainless steel tubing.

PLOT columns contain a porous layer of a solid adsorbent such as alumina, molecular sieve, or Porapak®. PLOT columns are well suited for the analysis of light fixed gases and other volatile compounds. A good example is the separation of krypton, neon, argon, oxygen, nitrogen, and xenon on a molecular sieve PLOT column as shown in Figure 6.4. PLOT columns represent a small (<5%) but important share of the GC column market.

OT COLUMN TUBING

Many types of column tubing including glass, copper, nylon, and stainless steel have been used; however, fused silica is by far the most popular today.

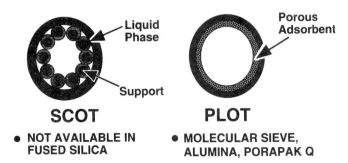

Fig. 6.3. Comparison of support coated open tubular (SCOT) column and porous layer open tubular (PLOT) column.

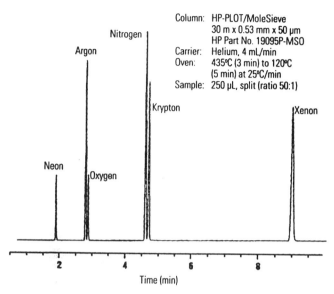

Fig. 6.4. Separation of fixed gases on a molecular sieve PLOT column. Copyright 1994 Hewlett-Packard Company. Reproduced with permission.

Stainless steel was introduced in the early days of capillary GC; however, it is not very efficient and is too active for highly reactive compounds such as steroids, amines, and free acids. Glass columns, unfortunately, are fragile.

Fused silica was introduced in 1979 [2] and today over 95% of all capillary columns sold are made of fused silica. Fused silica is flexible and easy to handle. It is also the most inert tubing material available and readily produces high resolution columns. The surface energy of fused silica matches well with the surface tension of silicon liquid phases. The silicon phases "wet" the tubing very well, resulting in very uniform thin films and very efficient columns.

Fused silica is made by the reaction of $SiCl_4$ and water vapor in a flame. The product, pure SiO_2, contains about 0.1% hydroxyl or silanol groups on the surface and less than 1 ppm of impurities (Na, K, Ca, etc.). The high purity of fused silica is responsible for its very inert chemical nature. A working temperature of about 1800°C is required to soften and draw fused silica into capillary dimensions. Fused silica columns are drawn on expensive sophisticated machinery using advanced fiber optics technology.

Fused silica has a high tensile strength and most chromatographic columns have a very thin wall, about 25 micrometers. This makes them flexible and easy to handle. The thin wall, however, is subject to rapid corrosion and breakage, even on exposure to normal laboratory atmospheres. Therefore, a thin protective sheath of polyimide is applied to the outside of the tubing as it emerges from the drawing oven. This polyimide coating, which darkens with age, protects the fused silica from atmospheric moisture. It is this polyimide coating that limits most fused silica columns to a maximum operating temperature of 360°C (short term 380°C). For higher column temperatures, stainless steel-clad fused silica is required.

ADVANTAGES OF OT COLUMNS

Table 6.1 explains why capillary columns are so popular. Since they are open tubes, there is little pressure drop across them; thus long lengths, such as 60 meters, can easily be used. Packed columns, on the other hand, are tightly packed with solid support, producing greater pressure drops and making long lengths impractical. A typical packed column length is two meters.

Capillary columns may be coated with a thin, uniform liquid phase because of fused silica's smooth, inert surface, which generates a high efficiency, typically 3,000 to 5,000 theoretical plates per meter. Packed columns, on the other hand, have thicker, often nonuniform films, and generate only 2,000 plates per meter. Thus, total plates available in long capillary columns range from 180,000 to 300,000, while packed columns typically generate only 4,000 plates and show much lower resolution.

Figure 6.5 shows chromatograms of the same sample on a packed and capillary column. On top is a packed-column separation of AROCHLOR 1260®, a commercial blend of polychlorinated biphenyl compounds. A two-meter column of two-millimeter, i.d. glass was used with an electron capture detector. This chromatogram exhibits a plate number of about 1,500, and we observe about 16 peaks with this sample.

The bottom chromatogram shows the same sample run on a 50-meter capillary column. Because capillary columns have relatively low capacities, the vaporized sample was split, so that only one part in 31 entered the column. The "split injection" technique allows a very small amount of sample to be injected rapidly. This chromatogram shows much better resolution, over 65 peaks, and a faster analysis—52 minutes compared to 80 minutes for the packed column. Arochlor® is obviously a very complex mixture, and even this high resolution capillary column with 150,000 plates did not resolve all of the peaks.

TABLE 6.1 Comparison of Capillary and Packed Columns

	Capillary	Packed
Length	60 meters	2 meters
Theoretical plates (N/m)	3,000–5,000	2000
Total plates length × N/m	180–300 K	4000

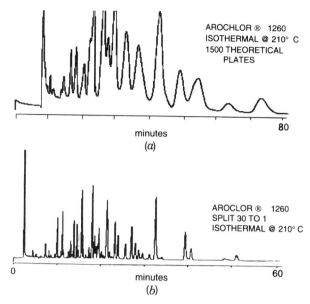

AROCHLOR ® 1260
ISOTHERMAL @ 210° C
1500 THEORETICAL
PLATES

minutes
80
(a)

AROCLOR ® 1260
SPLIT 30 TO 1
ISOTHERMAL @ 210° C

0 minutes 60
(b)

Fig. 6.5. Comparison of two separations of polychlorinated biphenyls (AROCHLOR)®: (*a*) packed column; (*b*) capillary column.

COLUMN SELECTION

The five critical parameters for capillary columns are: 1) internal diameter; 2) column length; 3) film thickness; 4) stationary phase composition; and 5) flow rate. Each will be discussed briefly.

Internal Column Diameter, i.d.

Internal column diameters for fused silica range from 100 to 530 micrometers (0.10–0.53 mm). Some glass capillaries have even larger internal diameters. One-hundred micrometer columns, row one of Table 6.2, have limited sample capacity, and are not well suited for trace analysis. Ease of operation is also limited because of the very limited sample capacity. These small i.d. columns have very good efficiency and produce fast analyses (see Fig. 6.6), but special sampling techniques and high-speed data systems are required to realize their full potential.

Many capillary columns have internal diameters of 250 or 320 micrometers, row two of Table 6.2. These i.d.'s represent the best compromise between resolution, speed, sample capacity, and ease of operation. These are the reference columns against which all other internal diameters are measured.

Five-hundred-and-thirty-micrometer or "widebore" columns, seen in row three of Table 6.2, show loss in resolution compared to analytical

TABLE 6.2   Effects of Column Diameter

Inside Diameter	Resolution	Speed	Capacity	Ease
100 μm	Very good	Very good	Fair	Fair
250 μm 320 μm	Good	Good	Good	Good
530 μm	Fair	Good	Very good	Very good

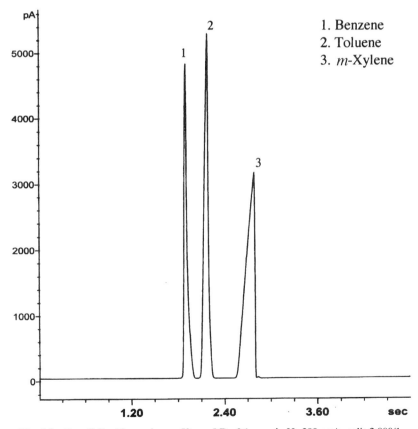

1. Benzene
2. Toluene
3. *m*-Xylene

Fig. 6.6.   Fast GC—10 m column, 50 μm I.D., 0.1 μm d_f, H$_2$ 200 cm/s, split 2,000/1.

capillary columns. This limitation is offset in most applications by their increased capacity and ease of operation. For example, direct on-column syringe injection is possible, often providing better quantitative results than packed columns. These widebore or "megabore" columns also show good speed of analysis.

Column Length

Plate number, N, is directly proportional to column length, L; the longer the column, the more theoretical plates, the better the separation. Resolution, R_s, however, is only proportional to the square root of column length. This means that if column length is doubled, plate number is doubled, but resolution only increases by the square root of two, or 41%.

Retention time, t_R, is also proportional to column length, so long columns can lead to slow analysis times. But when high resolution is critical, long columns are required. Referring to Table 6.3, columns that are 60 meters long are suggested for natural products such as flavors and fragrances—in fact, for any sample with more than 50 components. Remember, however, that analysis times will be long.

For fast analysis of simpler samples, short columns of perhaps ten meters should be used. Only moderate resolution is possible, but speed of analysis can be impressive. Figure 6.6 showed a fast analysis on a very short, one-meter, thin-film capillary column. A 50 mm i.d. column was used with a split ratio of 500/1. Note the baseline resolution of benzene, toluene, and *o*-xylene in under three seconds!

Medium column lengths of 25 or 30 meters are recommended for most applications. They provide a good compromise between resolution and speed of analysis.

Film Thickness

A standard film thickness of 0.25 μm is a reasonable starting point. It represents a compromise between the high resolution attainable with thin

TABLE 6.3 Column Length Recommendations

Column Length		Resolution	Speed
	Long (60–100 m)	High	Slow
	Short (5–10 m)	Moderate	Fast
	Medium (25–30 m)	Good compromise, good starting point	

films and the high capacity available with thick films. High capacity means that not only can larger sample quantities be accommodated, but usually the injection technique is also simpler.

With 0.25 μm films, practical operating temperatures can be used with minimal concern for column bleed, since column bleed is proportional to the amount of liquid phase in the column. Finally, with this film thickness, the column can be optimized for high speed using fast flow rates or high resolution using slower flow rates.

Thick films (1.0 μm or greater) are made possible today due to improved techniques in cross-linking liquid phases, and also to the more inert fused silica surface. Cross-linking techniques will be discussed later in this chapter. Such thick films show increased retention of sample components—essential for volatile compounds. In addition, their high capacity allows injection of larger samples; this can be important when mass spectrometers or Fourier transform-infrared spectrometers are to be used for subsequent analysis.

Decreased efficiency is one disadvantage of thick films. Thus, greater lengths may be required to compensate for their lower plate numbers. Also, higher operating temperatures are required to elute compounds from thick films. Higher temperatures, in turn, produce higher bleed rates and/or more noise. Also, since column bleed is proportional to the amount of liquid phase in the column, thick films do bleed more.

Figure 6.7 is a typical thick film application, the separation of natural gas components using a 50-meter column. The film thickness is five micrometers of polydimethylsiloxane, chemically bonded. Note the excellent resolution of methane, ethane, propane, and *n*-butane: peaks one, two, three, and four. This column is well suited for volatile compounds but should not be used for high molecular weight samples, as it would require excessively

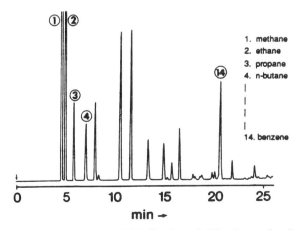

Fig. 6.7. Natural gas separation on a thick film (5 μm) OT column. Conditions: 50-m by 320 μm WCOT, CP-sil 8 CB, 40°C isothermal for 1 min., programmed 5°C/min to 200°C.

high temperatures and long analysis times, Note, for example, that benzene (peak 14) takes 20 minutes to elute even at 140°C.

The primary advantage of thin films, defined as less than 0.2 μm, is high efficiency and, therefore, higher resolution. Thus shorter columns can be used for many applications (refer back to Fig. 6.6). In addition, lower operating temperatures may be used, giving less column bleed.

Stationary Liquid Phases

Liquid phases for capillary columns are very similar to those used for packed columns. In both cases the liquid phase must show high selectivity, α, for the compounds of interest. In addition, they should be capable of operation at high temperatures with minimal column bleed. This is particularly important for sensitive detectors like FID, ECD, and MS which are used for trace analysis.

Table 6.4 lists the most commonly used liquid phases for both packed and capillary columns. Basically, there are two types of liquid phases in use today. One is siloxane polymers, of which OV-1, SE-30, DB-1 (100% methyl polysiloxane) and OV-17, OV-275, DB-1701, DB-710 (mixtures of methyl, phenyl, and cyano) polysiloxanes are the most popular. The other common liquid phase is a polyethylene glycol (Carbowax 20M, Superox® and DB-wax®).

Schematic structures of both a dimethyl polysiloxane and a polyethylene glycol liquid phase were given in Chapter 4. There is, however, one difference between packed column and capillary column liquid phases: capillary column phases are extensively cross-linked. By heating the freshly prepared capillary column at high temperatures (without column flow) the methyl groups form free radicals which readily cross-link to form a more stable, higher molecular weight gum phase. There is even some chemical bonding with the silanol groups on the fused silica surface. These cross-linked and chemically bonded phases are more temperature stable, last longer and can be cleaned by rinsing with solvents when cold. Most commercial capillary columns are cross-linked.

Carrier Gas and Flow Rate

Van Deemter plots were shown in Chapter 3, and they illustrate the effect of column flow rate on band broadening, H. There is an optimal flow rate for a minimum of band broadening. With packed columns, and also with thick film megabore columns, nitrogen is the carrier gas of choice since the van Deemter B term (longitudinal diffusion in the gas phase) dominates. Nitrogen being heavier than helium minimizes this B term and produces more efficiency.

In capillary columns, however, particularly those with thin films, hydrogen is the best carrier gas (refer to Figure 3.12). With capillary columns the efficiency (N) is usually more than sufficient and the emphasis is on

TABLE 6.4 Equivalent Stationary Phases

Supelco	Alltech	Chrompack	HP	J&W	Quadrex	Restek	SGE	Nonbonded/Packed Column Phases
SPB-1	AT-1	CP-Sil 5 CB	HP-1	DB-1	007-1	Rt_x-1	BP-1	SE-30, SP-2100, OV-1, OV-101
SPB-5	AT-5	CP-Sil 8 CB	HP-5	DB-5	007-2	Rt_x-5	BP-5	SE-54, SE-52, OV-73
SPB-20	—	—	—	—	007-7	Rt_x-20	—	OV-7
SPB-35	AT-35	—	—	—	007-11	RT_x-35	—	OV-11
SPB-1701	AT-1701	CP-Sil 19 CB	HP-1701	DB-1701	007-1701	Rt_x-1701	BP-10	OV-1701
Nukol	AT-1000	CP-Wax 58 CB	HP-FFAP	DB-FFAP	007-FFAP	Stabil wax-DA	DP-20	SP-1000, OV-351
Supelcowax 10	AT-Wax	CP-Wax 52 CB	HP-20 M	DB-WAX	007-CW	Stabil wax	BP-21	Carbowax 20M
SP-2330	—	CP-Sil 84	—	DB-23	007-23	Rt_x-2330	BPX-70	SP-2330, SP-2340, OV-275
SP-2380	—	CP-Sil 88	—	—	—	Rt_x-2340	—	
SP-2340	—	—	—	—	—	—	—	

Reprinted with permission of Supelco, Bellefonte, PA 16823 USA from their 1995 catalog.

speed. Thus, capillary columns are usually run at faster-than-optimal flow rates where the C_M term, mass transfer in the mobile phase, dominates. Hydrogen provides a much faster analysis with a minimal loss in efficiency because it allows faster diffusion in the mobile phase and minimizes the C_M term in the Golay equation. For example, refer back to the fast chromatogram in Figure 6.6 where hydrogen carrier gas was used at 5 times faster than the optimal flow rate. This high-speed analysis is not possible with packed columns or even thick film capillary columns.

CAPILLARY INLET SYSTEMS

Capillary columns have very strict requirements for sample injections: The injection profile should be very narrow (fast injections) and the quantity should be very small, usually less than 1 microgram. A typical 25-m capillary column contains about 10 mg of liquid phase, compared to 2 to 3 g for a 6-ft packed column. This explains why a very small sample should be injected; it is necessary to avoid "over-loading" the column.

Capillary peaks are usually very narrow, often having peak widths of a few seconds, so very fast injections are necessary to minimize the band broadening from a slow injection. There are numerous injection techniques used in capillary GC; in fact, entire textbooks have been written about the topic [3]; but here we will discuss only the most common techniques.

Split Injection

Split injection is the oldest, simplest, and easiest injection technique to use. The procedure involves injecting 1 μL of the sample by a standard syringe into a heated injection port that contains a deactivated glass liner. The sample is rapidly vaporized, and only a fraction, usually 1–2%, of the vapor enters the column (see Fig. 6.8). The rest of the vaporized sample and a large flow of carrier gas passes out through a split or purge valve.

There are several advantages to split injections. The technique is simple because the operator has only to control the split ratio by opening or closing the split (purge) valve. The sample amount introduced to the column is very small (and easily controlled), and the flow rate up to the split point is fast (the sum of both column and vent flow rates). The result is high-resolution separations. Another advantage is that "neat" samples can be introduced, usually by using a larger split ratio, so there is no need to dilute the sample. A final advantage is that "dirty" samples can be introduced by putting a plug of deactivated glass wool in the inlet liner to trap non-volatile compounds.

One disadvantage is that trace analysis is limited since only a fraction of the sample enters the column. Consequently, splitless or on-column injection techniques are recommended for trace analysis.

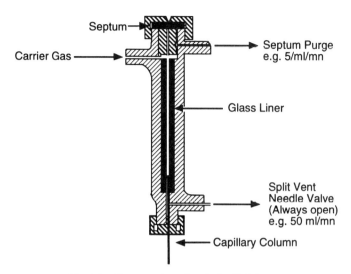

Fig. 6.8.   Cross section of typical split injector.

A second disadvantage is that the splitting process sometimes discriminates against high molecular weight solutes in the sample so that the sample entering the column is not respresentative of the sample injected.

Splitless Injection

Splitless injection uses the same hardware as split injection (Fig. 6.9), but the split valve is initially closed. The sample is diluted in a volatile solvent (like hexane or methanol) and 1 to 5 μL is injected in the heated injection port. The sample is vaporized and slowly (flow rate of about 1 mL/min) carried onto a cold column where both sample and solvent are condensed. After 45 seconds, the split valve is opened (flow rate of about 50 mL/min), and any residual vapors left in the injection port are rapidly swept out of the system. Septum purge is essential with splitless injections.

The column is now temperature programmed, and initially only the volatile solvent is vaporized and carried through the column. While this is happening, the sample analytes are being refocused into a narrow band in the residual solvent. At some later time, these analytes are vaporized by the hot column and chromatographed. High resolution of these higher boiling analytes is observed.

The big advantage of splitless injection is the improved sensitivity over split. Typically 20- to 50-fold more sample enters the column and the result is improved trace analysis for environmental, pharmaceutical, or biomedical samples.

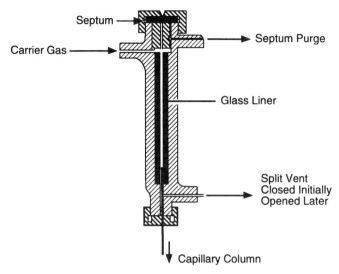

Fig. 6.9. Cross section of typical splitless injector.

Splitless has several disadvantages. It is time-consuming; you must start with a cold column; and you must temperature program. You must also dilute the sample with a volatile solvent, and optimize both the initial column temperature and the time of opening the split valve. Finally, splitless injection is not well-suited for volatile compounds. For good chromatography the first peaks of interest must have boiling points 30°C higher than the solvent.

Other Inlets

Three other types of capillary inlets are "direct injection," "on-column" and "cold on-column." Direct injection involves injecting a small sample (usually 1 μL or smaller) into a glass liner where the vapors are carried directly to the column. On-column means inserting the precisely aligned needle into the capillary column, usually a 0.53-mm-i.d. megabore, and making injections inside the column. Both of these techniques require thick film capillaries and wide-diameter columns with faster than normal flow rates (~10 mL/min). Even with these precautions the resolution is not as good as with split or splitless injection. The advantages can be better trace analysis and good quantitation.

Both high resolution and good quantitation result from cold on-column injections. A liquid sample is injected into either a cold inlet liner or a cold column. The cold injector is rapidly heated and the sample vaporized and carried through the column. Minimal sample decomposition is observed. For thermolabile compounds, cold on-column is the best injection tech-

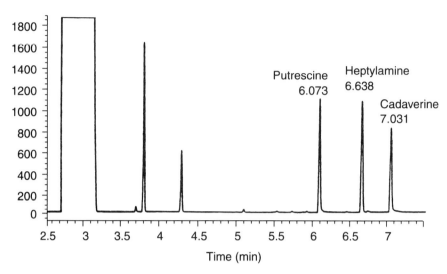

Fig. 6.10. Difficult amine separation using cold on-column injection. Reproduced from the *Journal of Chromatographic Science,* by permission of Preston Publications, a Division of Preston Industries, Inc.

nique. Unfortunately, these injectors are expensive accessories at this time and are not commonly used. One application, shown in Figure 6.10, is an amine separation for which cold on-column injection is optimal [4].

REFERENCES

1. Golay, M. J. E., *Gas Chromatography 1958* (*Amsterdam Symposium*), D. H. Desty, Ed., Butterworths, London, 1958, pp. 36–55 and 62–68.
2. Dandeneau, R. D., and Zerenner, E. H., *J. High Resolut. Chromatogr.,* **2,** 351–356 (1979).
3. Grob, K., *Classical Split and Splitless Injection in Capillary Gas Chromatography,* 3rd. ed., Heuthig, Heidelberg, 1993.
4. Bonilla, M., Enriquez, L. G., and McNair, H. M., *J. Chromatogr. Sci.,* **35,** 53 (1997).

7. Detectors

With a few exceptions, most of the detectors used in GC were invented specifically for this technique. The major exceptions are the thermal conductivity detector (TCD, or katharometer) that was preexisting as a gas analyzer when GC began, and the mass spectrometer (or mass selective detector, MSD) that was adapted to accept the large volumes and the fast scan rates needed for GC peaks. Most recently, other spectroscopic techniques like IR and atomic plasma emission have been coupled to the effluent from gas chromatographs, serving as GC detectors.

In total, there are probably over 60 detectors that have been used in GC. Many of the "invented" detectors are based on the formation of ions by one means or another, and of these, the flame ionization detector (FID) has become the most popular. The most common detectors are listed in Table 7.1; those that are highly selective are so designated in column two.

In one of the early books on detectors, David [1] discussed eight detectors in detail (see Table 7.1) and another dozen briefly, indicating that these 20 detectors were the most popular in the 1970s. More recently, Hill and McMinn [2] have edited a book describing the twelve detectors they feel are the most important in capillary GC. Scott's new book on chromatographic detectors [3] includes the FID, NPD, and photometric detectors in one section, the argon ionization types (including He ionization and ECD) in a second section, and katharometer types (including TCD, GADE, and radiometric) in a third. A few selective detectors are thoroughly described

TABLE 7.1   Common Commercially Available Detectors

Name	Selective*	References
1. Flame Ionization Detector (FID)	No	1, 2, 3, 5, 12, 13
2. Thermal Conductivity Detector (TCD) (Katharometer)	No	1, 3, 5
3. Electron Capture Detector (ECD)	X	1, 2, 3, 5
Other ionization type detectors		
4. Nitrogen/Phosphorous Detector (NPD); Alkali Flame Ionization Detector (AFID); Thermionic Ionization Detector (TID)	N, P, X	1, 2, 3, 5
5. Photoionization Detector (PID); Discharge Ionization Detector (DID)	Aromatics	2, 5
6. Helium Ionization Detector (HID)	No	1, 2, 3, 5
Emission type detectors		
7. Flame Photometric Detector (FPD)	S, P	1, 2, 3, 5
8. Plasma Atomic Emission (AED)	Metals, X, C, O	2, 4, 5
Electrochemical detectors		
9. Hall Electrolytic Conductivity (HECD)	S, N, X	1, 2, 5
Other types of detector		
10. Chemiluminescent	S	2, 4
11. Gas Density Detector (GADE)	No	1, 3, 5
12. Radioactivity Detector	^{3}H, ^{14}C	3
13. Mass Spectrometer (MS or MSD)	Yes	2, 4
14. Fourier Transform Infrared (FTIR)	Yes	2, 5

* X = Halogen

in a book edited by Sievers [4]; it is highly specialized and devoted mainly to elemental analysis. These and other references in Table 7.1 can be consulted for more detailed information.

In this chapter, the FID, TCD, and electron capture detector (ECD) will be featured since they are the three most widely used detectors. A few of the others from Table 7.1 will be described briefly; in addition, the combination of GC and MS is so important that it is treated separately in Chapter 10. First, however, some classifications and common properties will be discussed in order to provide a comprehensive framework for this chapter.

CLASSIFICATION OF DETECTORS

Of the five classification systems listed below, the three most important ones are discussed in this section; the other two are fairly obvious. Included in the list are the classifications for the FID, TCD, and ECD.

Classification of GC Detectors

1. Concentration	vs.	Mass flow rate
TCD ECD		FID
2. Selective	vs.	Universal
ECD (FID)		TCD
3. Destructive	vs.	Nondestructive
FID		TCD ECD
4. Bulk property	vs.	Specific property
TCD ECD		FID
5. Analog	vs.	Digital
FID TCD ECD		

Concentration vs. Mass Flow Rate

This classification system distinguishes between those detectors that measure the *concentration* of the analyte in the carrier gas compared to those that directly measure the absolute *amount* of analyte irrespective of the volume of carrier gas. Note in the first example in the list above that the TCD and ECD are concentration types and the FID is a mass flow rate type. One consequence of this difference is that peak areas and peak heights are affected by changes in carrier gas flow rate.

To understand the reason for this difference in detector type, consider the effect on a TCD signal if the flow is completely stopped. The detector cell remains filled with a given concentration of analyte and its thermal conductivity continues to be measured at a constant level. However, for a mass flow rate detector like the FID in which the signal arises from a burning of the sample, a complete stop in the flow rate will cause the delivery of the analyte to the detector to stop and the signal will drop to zero.

Figure 7.1 shows the effect of decreased flow rate on the peaks from the two types of detector: for the concentration type, the area increases and the height is unchanged; for the mass flow rate type, the peak height is decreased and the area is unchanged. Consequently, quantitative data acquired at different flow rates will be affected. While these variations can be eliminated by using standards or electronic flow regulators, it is often the case that flow rates will change during an individual run if the chromatograph is being operated at a constant pressure during programmed temperature operation (as, for example, following split/splitless injection sampling). For this reason, operation at constant flow may be necessary for quantitative analysis in programmed temperature GC, and it is easily achieved today with the use of electronic flow controllers. Fortunately, if one is performing a quantitative analysis by programmed temperature at constant pressure with a FID, peak *areas* are unaffected.

This difference in performance has two other consequences. First of all, it is difficult to compare the sensitivities of these two types of detectors

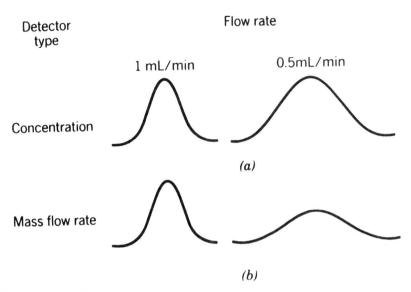

Fig. 7.1. Effect of flow rate on peak sizes for the two types of detector: (*a*) concentration and (*b*) mass flow rate. From Miller, J. M., *Chromatography: Concepts and Contrasts,* John Wiley & Sons, Inc., New York, 1987, p. 91. Reproduced courtesy of John Wiley & Sons, Inc.

because their signals have different units; the better comparison is between minimum detectable quantity which has the units of mass for both types. And second, valid comparisons between detector types requires the specification of the flow rate and concentration.

Operation of all detectors is optimized when their internal volumes are small, since band broadening is thereby minimized. However, concentration detectors have a cell volume in which detection occurs and the magnitude of that volume has special importance. Suppose the cell volume of a concentration detector is so large that the entire sample could be contained in one cell volume. The shape of the resulting peak would be badly broadened and distorted.

Estimates can be made of ideal cell volume requirements, since the width of a peak can be expressed in volume units (the base width, 4σ, where the x-axis is in mL units). A narrow peak from a capillary column might have a width as small as 1 second, which represents a volume of 0.017 mL (17 μL) at a flow rate of 1 mL/min. If the detector volume were the same or larger, the entire peak could be contained in it at one time and the peak would be very broad. An ideal detector for this situation should have a significantly smaller volume, say 2 μL. When this is not possible, make-up gas can be added to the column effluent to sweep the sample through the detector more quickly. This remedy will be helpful for mass flow rate detectors but less so for concentration detectors. In the latter case, the make-up gas dilutes the sample, lowering the concentration as well as the resulting signal—not a satisfactory solution in some cases. Consequently, concentration detectors must have very small volumes if they are to be used successfully for capillary GC. Make-up gas may also be used with them, but at the risk of decreasing the signal.

Selective vs. Universal

This detector category refers to the number or percentage of analytes that can be detected by a given system. A universal detector theoretically detects all solutes, while the selective type responds to particular types or classes of compounds. There are differing degrees of selectivity; the FID is not very selective and detects all organic compounds while the ECD is very selective and detects only very electronegative species, like halogen-containing pesticides.

Both types of detectors have advantages. The universal detectors are used when one wants to be sure all eluted solutes are detected. This is important for qualitative screening of new samples whose composition is not known. On the other hand, a selective detector that has enhanced sensitivity for a small class of compounds can provide trace analysis for that class even in the presence of other compounds of higher concentration. It can simplify a complex chromatogram by detecting only a few of the compounds present, and selectively "ignoring" the rest. As an example, the FPD can selectively detect only sulfur-containing compounds in a forest of hydrocarbon peaks in a gasoline or jet fuel sample.

Destructive vs. Nondestructive

Nondestructive-type detectors are necessary if the separated analytes are to be reclaimed for further analysis, as, for example, when identifications are to be performed using auxiliary instruments. One way to utilize destructive detectors in such a situation would be to split the effluent stream and send only part of it to the detector, collecting the rest for analysis.

DETECTOR CHARACTERISTICS

The most important detector characteristic is the signal it produces, of course, but two other important characteristics are noise and time constant. The latter two will be discussed first to provide a background for the discussion about the signal.

Noise

Noise is the signal produced by a detector in the absence of a sample. It is also called the background and it appears on the baseline. Usually it is given in the same units as the normal detector signal. Ideally, the baseline should not show any noise, but random fluctuations do arise from the electronic components from which the amplifiers are made, from stray signals in the environment, and from contamination and leaks. Circuit design can eliminate some noise; shielding and grounding can isolate the detector from the environment; and sample pretreatment and pure chromatographic gases can eliminate some noise from contamination.

The definition of noise used by ASTM (formerly the American Society for Testing and Materials) is depicted in Figure 7.2. The two parallel lines drawn between the peak-to-peak maxima and minima enclose the noise, given in mV in this example. In addition, the figure shows a long-term noise or *drift* occurring over a period of 30 minutes. If at all possible, the sources of the noise and drift should be found and eliminated or minimized because they restrict the minimum signal that can be detected. Some suggestions for reducing noise can be found in reference 6.

The ratio of the signal to the noise is a convenient characteristic of detector performance. It conveys more information about the lower limit of detection than does the noise alone. Commonly, the smallest signal that can be attributed to an analyte is one whose signal-to-noise ratio or S/N, is 2 or more. An S/N ratio of 2 is shown in Figure 7.3; it can be seen that this is certainly a minimum value for distinguishing a peak from the background noise. Sharp spikes which exceed an S/N of 2 should not be interpreted as peaks as these often arise from contamination and represent a different type of detector instability.

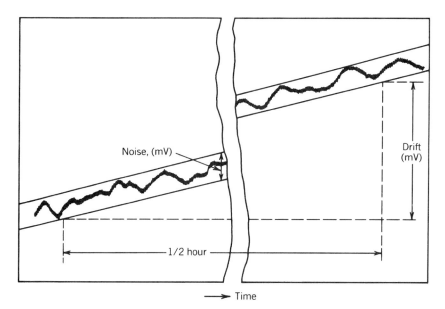

Fig. 7.2. Example of noise and drift in a TCD. Copyright ASTM. Reprinted with permission. From Miller, J. M., *Chromatography: Concepts and Contrasts,* John Wiley & Sons, Inc., New York, 1987, p. 93. Reproduced courtesy of John Wiley & Sons, Inc.

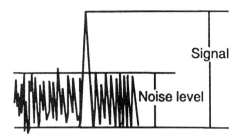

Fig. 7.3. Illustration of a signal-to-noise ratio (S/N) of 2. Reprinted from Grant, D. W., *Capillary Gas Chromatography,* Wiley, 1996. Copyright John Wiley & Sons, Inc. Reproduced with permission.

Time Constant

The time constant, τ, is a measure of the speed of response of a detector. Specifically, it is the time (in seconds or milliseconds) a detector takes to respond to 63.2% of a sudden change in signal as shown in Figure 7.4. The full response (actually 98% of full response) takes four time constants and is referred to as the *response time*[a]. One of these two parameters should be specified for a detector.

Figure 7.5 shows the effect of increasingly longer time constants that distort the shape of a chromatographic peak. The deleterious effects on chromatographic peaks are the changes in retention time (peak position in the chromatogram) and on peak width, both of which get larger as the time constant increases. The area, however, is unaffected; quantitative measurements based on area will remain accurate while only those based on peak height will be in error.

A typical recommendation [7] is that the time constant should be less than 10% of the peak width at half height, w_h. Thus, a peak width of 50 μL at a flow of 1 mL/min corresponds to a time constant of 0.3 seconds. This is the order of magnitude required for most chromatographic detectors and their associated data systems. Remember also that the overall time constant for the entire system is limited by the largest value for any of the individual components.

Large time constants do have the advantage of decreasing the short-term noise from a detector. This effect is sometimes called *damping*. The temptation to decrease one's chromatographic noise and improve one's chromatograms by increasing the time constant must be avoided. Valuable information can be lost when the data system does not faithfully record all the available information, including noise.

---

[a] Unfortunately, some workers define response time as 2.2 time constants (not 4.0) that corresponds to 90% (not 98%) of full scale deflection; others define a *rise time* as the time required for a signal to rise from 10% to 90%. This lack of consistency can even be found in some ASTM specifications.

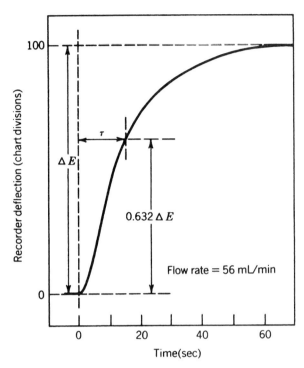

Fig. 7.4. Illustration of the definition of response time. Copyright ASTM. Reprinted with permission. From Miller, J. M., *Chromatography: Concepts and Contrasts,* John Wiley & Sons, Inc., New York, 1987, p. 94. Reproduced courtesy of John Wiley & Sons, Inc.

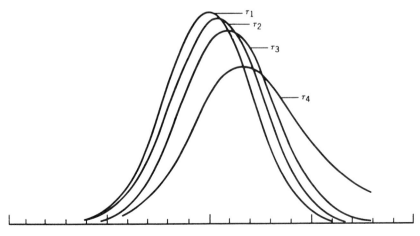

Fig. 7.5. Effect of detector time constant on peak characteristics; $\tau_1 < \tau_2 < \tau_3 < \tau_4$. From Miller, J. M., *Chromatography: Concepts and Contrasts,* John Wiley & Sons, Inc., New York, 1987, p. 95. Reproduced courtesy of John Wiley & Sons, Inc.

Signal

The detector output or signal is of special interest when an analyte is being detected. The magnitude of this signal (peak height or peak area) is proportional to the amount of analyte and is the basis for quantitative analysis. Its characteristics are very important because quantitative analysis is an important application for GC. The signal specifications to be defined are sensitivity, minimum detectability, linear range, and dynamic range.

Sensitivity

The detector sensitivity, S, is equal to the signal output per unit concentration or per unit mass of an analyte in the carrier gas. The units of sensitivity are based on area measurements of the peaks and differ for the two main detector classifications, concentration and mass flow rate [8].

For a concentration type detector, the sensitivity is calculated per unit *concentration* of the analyte in the mobile gas phase,

$$S = \frac{AF_c}{W} = \frac{E}{C} \tag{1}$$

where A is the integrated peak area (in units like mV/min), E is the peak height (in mV), C is the concentration of the analyte in the carrier gas (in mg/mL), W is the mass of the analyte present (in mg), and F_c is the carrier gas flow rate (corrected[b]) in mL/min. The resulting dimensions for the sensitivity of a concentration detector are mV mL/mg.

For a mass flow rate type detector, the sensitivity is calculated per unit *mass* of the analyte in the mobile gas phase,

$$S = \frac{A}{W} = \frac{E}{M} \tag{2}$$

where M is the mass flow rate of the analyte entering the detector (in mg/sec), W is the mass of the analyte (in mg), the peak area is in ampere-sec, and the peak height is in amperes. In this case, the dimensions for sensitivity are ampere-sec/mg or coulomb/mg. As noted earlier, the differences in the units of sensitivity between the two types of detector makes comparisons of the sensitivities difficult.

Figure 7.6 shows a plot of detector signal versus concentration for a TCD, a concentration type detector. The slope of this line is the detector sensitivity according to equation 1. A more sensitive detector would have a greater slope and vice versa. Because the range of sample concentrations often extends over several orders of magnitude, this plot is often made on a log-log basis to cover a wider range on a single graph.

[b] For the definition of *corrected* flow rate, see Chapter 2.

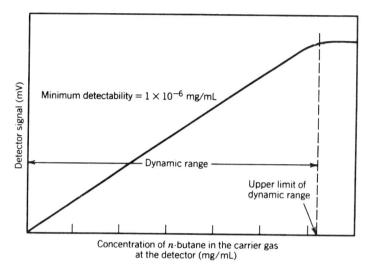

Fig. 7.6. Illustration of the definition of dynamic range for a TCD. Copyright ASTM. Reprinted with permission. From Miller, J. M., *Chromatography: Concepts and Contrasts,* John Wiley & Sons, Inc., New York, 1987, p. 96. Reproduced courtesy of John Wiley & Sons, Inc.

As shown at the upper end of the graph, linearity is lost and eventually the signal fails to increase with increased concentration. These phenomena will be discussed later in the section on linearity.

Minimum Detectability

The lowest point on Figure 7.6, representing the lower limit that can be detected, has been called by a variety of names such as minimum detectable quantity (MDQ), limit of detection (LOD), and detectivity. The IUPAC [8] has defined the *minimum detectability, D,* as,

$$D = \frac{2N}{S} \tag{3}$$

where N is the noise level and S is the sensitivity as just defined. Note that the numerator is multiplied by 2 in accordance with the definition discussed earlier that a detectable signal should be twice the noise level. The units of detectability are mg/mL for a concentration type detector and mg/sec for a mass flow rate type.

If the minimum detectability is multiplied by the peak width of the analyte peak being measured, and if the appropriate units are used, the value that results has the units of mg and represents the minimum mass that can be detected chromatographically, allowing for the dilution of the sample that results from the process. Some call this value the MDQ. As such, it is a convenient measure to compare detection limits between detectors of different types.

A related term is the limit of quantitation (LOQ) which should be above the LOD. For example, the ACS guidelines on environmental analysis [9] specify that the LOD should be three times the S/N and the LOQ ten times the S/N. The definitions of the USP are similar and also state that the LOQ should be no less than two times the LOD [10]. Other agencies may have other guidelines, but all are concerned with the same need to specify detection and quantitation limits, and the relationship between them. They are not the same.

Linear Range

The straight line in Figure 7.6 curved off and became nonlinear at high concentrations. It becomes necessary to establish the upper limit of linearity in order to measure the linear range. Since Figure 7.6 is often plotted on a log-log scale, the deviations from linearity are minimized and the curve is not a good one to use to show deviations. A better plot is one of *sensitivity* versus concentration as shown in Figure 7.7. Here the analyte concentration can be on a log scale to get a large range while the y-axis (sensitivity) can be linear. According to the ASTM specification, the upper limit of linearity is the analyte concentration corresponding to a sensitivity equal to 95% of the maximum measured sensitivity. The upper dashed line in the figure is drawn through the point representing the maximum sensitivity, and the lower dashed line is 0.95 of that value.

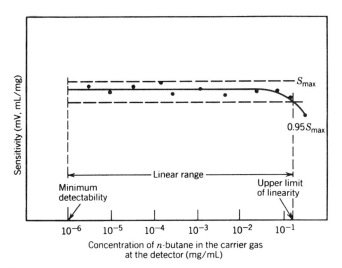

Fig. 7.7. Example of a linear plot of a TCD. Copyright ASTM. Reprinted with permission. From Miller, J. M., *Chromatography: Concepts and Contrasts,* John Wiley & Sons, Inc., New York, 1987, p. 97. Reproduced courtesy of John Wiley & Sons, Inc.

Having established both ends of the linear range, the minimum detectivity and the upper limit, the linear range is defined as their quotient:

$$\text{Linear range} = \frac{\text{Upper Limit}}{\text{Lower Limit}} \tag{4}$$

Since both terms are measured in the same units, the linear range is dimensionless. Obviously, a large value is desired for this parameter.

Linear range should not be confused with *dynamic range* which was indicated on Figure 7.6 as terminating at the point at which the curve levels off and shows no more increase in signal with increasing concentration. The upper limit of the dynamic range will be higher than the upper limit of the linear range and it represents the upper concentration at which the detector can be used.

Summary

The specification and selection of these detector characteristics is very important, especially for quantitative analysis (see Chapter 8). A good discussion about choosing the right detector settings, summarizing much of the material from this section has been published recently by Hinshaw [11].

FLAME IONIZATION DETECTOR (FID)

The FID is the most widely used GC detector, and is an example of the ionization detectors invented specifically for GC. The column effluent is burned in a small oxy-hydrogen flame producing some ions in the process. These ions are collected and form a small current that becomes the signal. When no sample is being burned, there should be little ionization, the small current (10^{-14} a) arising from impurities in the hydrogen and air supplies. Thus, the FID is a specific property-type detector with characteristic high sensitivity.

A typical FID design is shown in Figure 7.8. The column effluent is mixed with hydrogen and led to a small burner tip that is surrounded by a high flow of air to support combustion. An igniter is provided for remote lighting of the flame. The collector electrode is biased about $+300$ V relative to the flame tip and the collected current is amplified by a high impedance circuit. Since water is produced in the combustion process, the detector must be heated to at least 125°C to prevent condensation of water and high boiling samples. Most FIDs are run at 250°C or hotter.

The exact mechanism of flame ionization is still not known. Sternberg et al. [12] presented the early theories and Sevcik et al. [13] have presented a more recent discussion. The FID responds to all organic compounds that burn in the oxy-hydrogen flame. The signal is approximately proportional to the carbon content, giving rise to the so-called *equal per carbon* rule. A

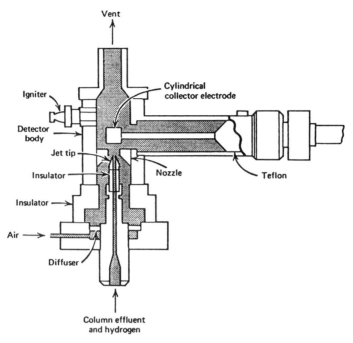

Vent

Igniter

Cylindrical
collector electrode

Detector
body

Jet tip

Insulator

Nozzle

Teflon

Insulator

Air

Diffuser

Column effluent
and hydrogen

Fig. 7.8. Schematic of an FID. Courtesy of Perkin-Elmer.

recent publication seems to confirm that the reason for this constant response factor is due to the conversion of all carbon atoms in an organic
solute to methane in the FID combustion process [14]. Thus, all hydrocarbons should exhibit the same response, per carbon atom. When hetero-
atoms like oxygen or nitrogen are present, however, the factor decreases.
Relative response values are often tabulated as *effective carbon numbers,*
ECN; for example, methane has a value of 1.0, ethane, 2.0, etc. Table 7.2
lists experimental and theoretical ECN values for some simple organic
compounds [15]. Clearly response factors are necessary for good quantitative analysis (see Chapter 8 for some weight response values).

For efficient operation, the gases (hydrogen and air) must be pure and
free of organic material that would increase the background ionization.
Their flow rates need to be optimized for the particular detector design
(and to a lesser extent, the particular analyte). As shown in Figure 7.9, the
flow rate of hydrogen goes through a maximum sensitivity for each carrier
gas flow rate, the optimum occurring at about the column flow rate. For
open tubular columns that have flows around 1 mL/min, make-up gas is
added to the carrier gas to bring the total up to about 30 mL/min.

Hydrogen can be used as the carrier gas, but changes in gas flows (a
separate source of hydrogen is still required) and detector designs are
required [16] in addition to the safety precautions that must be taken.

TABLE 7.2   **FID Effective Carbon Numbers (relative to heptane) [15]***

Compound	ECN	Theoretical ECN
Acetylene	1.95	2
Ethylene	2.00	2
Hexene	5.82	6
Methanol	0.52	0.5
Ethanol	1.48	1.5
n-Propanol	2.52	2.5
i-Propanol	2.24	2.5
n-Butanol	3.42	3.5
Amyl alcohol	4.37	4.5
Butanal	3.12	3
Heptanal	6.14	6
Octanal	6.99	7
Capric aldehyde	8.73	9
Acetic acid	1.01	1
Propionic acid	2.07	2
Butyric acid	2.95	3
Hexanoic acid	5.11	5
Heptanoic acid	5.55	6
Octanoic acid	6.55	7
Methyl acetate	1.04	2
Ethyl acetate	2.33	3
i-Propyl acetate	3.52	4
n-Butyl acetate	4.46	5
Acetone	2.00	2
Methyl ethyl ketone	3.07	3
Methyl i-butyl ketone	4.97	5
Ethyl butyl ketone	5.66	6
Di-i-butyl ketone	7.15	8
Ethyl amyl ketone	7.16	7
Cyclohexanone	4.94	5

* Reproduced from the *Journal of Chromatographic Science* by permission of Preston Publications, A Division of Preston Industries, Inc.

The flow rate of air is much less critical, and a value of 300 to 400 mL/min is sufficient for most detectors, as shown in Figure 7.10.

Compounds not containing organic carbon do not burn and are not detected. The most important ones are listed in Table 7.3. Most significant among those listed is water, a compound that often produces badly tailed peaks. The absence of a peak for water permits the FID to be used for analysis of samples that contain water since it does not interfere in the chromatogram[c]. Typical applications include organic contaminants in water, wine and other alcoholic beverages, and food products.

The list on page 116 summarizes the characteristics of the FID. Its advantages are good sensitivity, a large linearity, simplicity, ruggedness, and adaptability to all sizes of columns.

[c] Water may hydrolyze some of the polyester and polyethylene glycol liquid phases but it has little to no effect on silicone polymers.

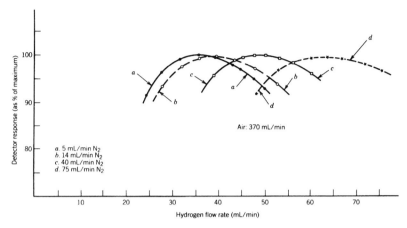

Fig. 7.9. Effect of hydrogen flow rate on FID response. Courtesy of Perkin-Elmer. From Miller, J. M., *Chromatography: Concepts and Contrasts,* John Wiley & Sons, Inc., New York, 1987, p. 128. Reproduced courtesy of John Wiley & Sons, Inc.

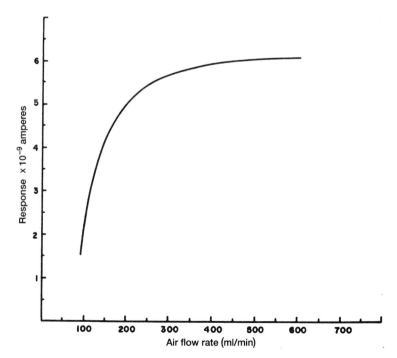

Fig. 7.10. Effect of air flow rate on FID response. Courtesy of Perkin-Elmer. From Miller, J. M., *Chromatography: Concepts and Contrasts,* John Wiley & Sons, Inc., New York, 1987, p. 129. Reproduced courtesy of John Wiley & Sons, Inc.

Flame Ionization Detector (FID) Characteristics

1. MDQ—10^{-11}g (~50 ppb)
2. Response—organic compounds only, no fixed gases or water
3. Linearity—10^6—excellent
4. Stability—excellent, little effect of flow or temperature changes
5. Temperature limit—400°C
6. Carrier gas—nitrogen or helium

THERMAL CONDUCTIVITY DETECTOR (TCD)

Nearly all of the early GC instruments were equipped with thermal conductivity detectors. They have remained popular, particularly for packed columns and inorganic analytes like H_2O, CO, CO_2, and H_2 (see Chapter 5).

The TCD is a differential detector that measures the thermal conductivity of the analyte in carrier gas, compared to the thermal conductivity of pure carrier gas. In a conventional detector at least two cell cavities are required, although a cell with four cavities is more common. The cavities are drilled into a metal block (usually stainless steel) and each contains a resistance wire or filament (so-called *hot wires*). The filaments are either mounted on holders, as shown in Figure 7.11, or are held concentrically in the cylindrical cavity, a design that permits the cell volume to be minimized. They are made of tungsten or a tungsten–rhenium alloy (so-called WX filaments) of high resistance.

The filaments are incorporated into a Wheatstone Bridge circuit, the classic method for measuring resistance (Fig. 7.12). A DC current is passed through them to heat them above the temperature of the cell block, creating a temperature differential. With pure carrier gas passing over all four elements, the bridge circuit is balanced with a "zero" control. When an analyte elutes, the thermal conductivity of the gas mixture in the two sample cavities is decreased, their filament temperatures increase slightly, causing the resistance of the filaments to increase greatly, and the bridge becomes unbalanced—that is a voltage develops across opposite corners of the

TABLE 7.3 Compounds Giving Little Or No Response In the Flame Ionization Detector

He	CS_2	NH_3
Ar	COS	CO
Kr	H_2S	CO_2
Ne	SO_2	H_2O
Xe	NO	$SiCl_4$
O_2	N_2O	$SiHCl_3$
N_2	NO_2	SiF_4

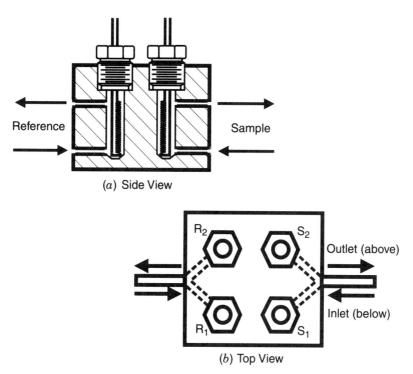

Fig. 7.11. Typical TCD cell, 4-filament. (*a*) Side view; (*b*) top view. Courtesy of the Gow-Mac Instrument Co., Bethlehem, PA. U.S.A.

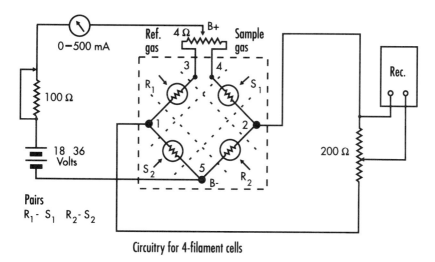

Fig. 7.12. Wheatstone bridge circuit for a 4-filament TCD. Courtesy of the Gow-Mac Instrument Co., Bethlehem, PA. U.S.A.

bridge. That voltage is dropped across a voltage divider (the so-called *attenuator*) and then all or part of it is fed to a recorder, integrator, or other data system. After the analyte is fully eluted, the thermal conductivity in the sample cavities returns to its former value and the bridge returns to balance.

The larger the heating current applied to the filaments, the greater the temperature differential and the greater the sensitivity. However, high filament temperatures also result in shorter filament life because small impurities of oxygen readily oxidize the tungsten wires, ultimately causing them to burn out. For this reason, the GC system must be free from leaks and operated with oxygen-free carrier gas.

The Wheatstone Bridge can be operated at constant voltage or constant current, but a more elaborate circuit can be used to maintain constant filament *temperature*. Thus, the detector controls may specify setting a current, a voltage, a temperature, or a temperature difference (ΔT), depending on the particular type of control. Controlling the filament temperature to keep it constant amounts to nulling the Bridge, unlike the simpler circuit which directly measures the bridge unbalance. Nulling provides a larger linear range, greater amplification, lower detection limits, and less noise [17].

As noted earlier in this chapter, a small cell volume is desirable for faithful reproduction of peak shapes and greater sensitivity. Typically, TCD cells have volumes around 140 μL which are very good for packed columns or wide bore capillaries. Their use with narrow capillaries has not become routine, but cells are available with volumes down to 20 μL and several studies have shown that good chromatograms can be obtained in some cases [18, 19]. Make-up gas is usually required when capillary columns are used with TCDs. An extremely small cell has been made by etching a nL volume on a silica chip for a micro-GC instrument. Another manufacturer uses a small volume (5 μL) single-cell TCD; in its operation the two gas streams (sample and reference) are passed alternately through the cell at a frequency of 10 times per second [5].

The carrier gas used with the TCD must have a thermal conductivity (TC) that is very different from the samples to be analyzed, so the most commonly used gases are helium and hydrogen which have the highest TC values [20]. It can be seen from the relative values listed in Table 7.4 that all other gases as well as liquids and solids have much smaller TC values. If nitrogen is used as a carrier gas, one can expect to get unusual peak shapes, often in the shape of a W due to partial peak inversion [21]. The same effect occurs if one attempts to analyze hydrogen using helium as the carrier gas [22].

Table 7.4 also contains some experimental relative response values for the samples listed. Although TCD response does not correlate directly with TC values, it is obvious the calibration factors are necessary for quantitative analysis, the same as for the FID.

A summary of TCD characteristics is given in the following list. TCD is a rugged, universal detector with moderate sensitivity.

Thermal Conductivity Detector (TCD) Characteristics

1. MDQ—10^{-9}g (~10 ppm)
2. Response—all compounds
3. Linearity—10^4
4. Stability—good
5. Carrier gas—helium
6. Temperature limit—400°C

ELECTRON CAPTURE DETECTOR (ECD)

The invention of the ECD (for GC) is generally attributed to Lovelock, based on his publication in 1961 [23]. It is a selective detector that provides very high sensitivity for those compounds that "capture electrons." These compounds include halogenated materials like pesticides and, consequently, one of its primary uses is in pesticide residue analysis.

TABLE 7.4 Thermal Conductivities and TCD Response Values for Selected Compounds [20]*

Compound	Thermal Conductivity[a]	RMR[b]
Carrier Gases		
Argon	12.5	—
Carbon dioxide	12.7	—
Helium	100.0	—
Hydrogen	128.0	—
Nitrogen	18.0	—
Samples		
Ethane	17.5	51
n-Butane	13.5	85
n-Nonane	10.8	177
i-Butane	14.0	82
Cyclohexane	10.1	114
Benzene	9.9	100
Acetone	9.6	86
Ethanol	12.7	72
Chloroform	6.0	108
Methyl iodide	4.6	96
Ethyl acetate	9.9	111

* Reproduced from the *Journal of Chromatographic Science* by permission of Preston Publications, A Division of Preston Industries, Inc.

[a] Relative to He = 100.

[b] Relative molar response in helium. Standard: benzene = 100.

It is an ionization-type detector, but unlike most detectors of this class, samples are detected by causing a *decrease* in the level of ionization. When no analytes are present, the radioactive ^{63}Ni emits beta particles as shown in equation (5):

$$^{63}Ni \rightarrow \beta^-$$ (5)

These negatively charged particles collide with the nitrogen carrier gas and produce more electrons (equation 6):

$$\beta^- + N_2 \rightarrow 2e^- + N_2^+$$ (6)

The electrons formed by this combined process result in a high standing current (about 10^{-8} a) when collected by a positive electrode. When an electronegative analyte is eluted from the column and enters the detector, it captures some of the free electrons and the standing current is decreased giving a negative peak:

$$A + e^- \rightarrow A^-$$ (7)

The negative ions formed have slower mobilities than the free electrons and are not collected by the anode.

The mathematical relationship for this process is similar to Beers Law (used to describe the absorption process for electromagnetic radiation). Thus, the extent of the absorption or capture is proportional to the concentration of the analyte. Some relative response values are given in Table 7.5 [24]; the high selectivity for halogenated materials can be seen from these data.

The carrier gas used for the ECD can be very pure nitrogen (as indicated in the mechanism presented) or a mixture of 5% methane in argon. When used with a capillary column some make-up gas is usually needed, and it is convenient to use inexpensive nitrogen as make-up and helium as the carrier gas.

A schematic of a typical ECD is shown in Figure 7.13. ^{63}Ni is shown as the beta emitter although tritium has also been used; nickel is usually preferred because it can be used at a higher temperature (up to 400°C) and it has a lower activity (and is safer).

It has been shown that improved performance is obtained if the applied voltage is pulsed rather than applied continuously. A square-wave pulse of around -50 V is applied at a frequency that maintains a constant current whether or not an analyte is in the cell; consequently the pulse frequency is higher when an analyte is present. The pulsed ECD has a lower MDQ and consequently a larger linear range. An example of pesticide residue analysis at the femtogram level is shown in Figure 7.14.

TABLE 7.5 Relative ECD Molar Responses [24]*

M	ECD Response[a]
CH_3Cl	1.4
CH_2Cl_2	3.5
$CHCl_3$	420
CCl_4	10,000
CH_3CH_2Cl	1.9
CH_2ClCH_2Cl	4.2
$CH_3CHClCH_3$	1.8
$(CH_3)CCl$	1.5
$CH_2 = CHCl$	0.0062
$CH_2 = CCl_2$	17
trans-$CHCl = CHCl$	1.5
cis-$CHCl = CHCl$	1.1
$CHCl = CCl_2$	460
$CCl_2 = CCl_2$	3,600
$Ph - Cl$[b]	0.026
$Ph - CH_2Cl$[b]	38
CF_3Cl	6.3
CHF_2Cl	1.8
CF_2Cl_2	160
$CFCl_3$	4,000

* Reproduced with permission from *Detectors for Capillary Chromatography*, by Grimsrud. Copyright John Wiley & Sons, Inc.

[a] Relative molar responses measured at 250°C in nitrogen detector gas using a Varian 3700 GC/CC-ECD.

[b] Ph = phenyl.

One drawback of the ECD is the necessity to use a radioactive source which may require a license or at least regular radiological testing. A new innovation is an ECD operated with a pulsed discharge (PDD) so that it does not require a radioactive source [25]. This detector is commercially available and can also be operated as a helium ionization detector under different conditions.

The ECD is one of the most easily contaminated detectors and is adversely affected by oxygen and water. Ultrapure, dry gases, freedom from leaks, and clean samples are necessary. Evidence of contamination is usually a noisy baseline or peaks that have small negative dips before and after each peak. Cleaning can sometimes be accomplished by operation with hydrogen carrier gas at a high temperature to burn off impurities, but dismantling is often required.

The list on page 123 gives the characteristics of the ECD. In summary, it is a sensitive and selective detector for halogenated materials but one which is easily contaminated and more prone to problems.

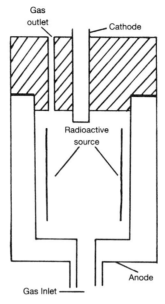

Fig. 7.13. Schematic of an ECD.

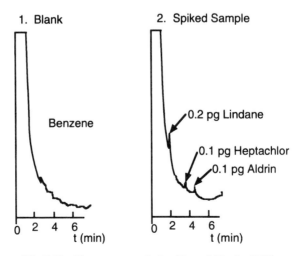

Fig. 7.14. Femtogram analysis of insecticides by ECD.

Summary of ECD Characteristics

1. MDQ—10^{-9} to 10^{-12}g
2. Response—very selective
3. Linearity—10^3 to 10^4
4. Stability—fair

OTHER DETECTORS

Table 7.1 listed the major detectors that are commercially available and in common use. Brief descriptions of a few of them are included here, and Figure 7.15 shows a comparison of the linear ranges of many of them.

Nitrogen Phosphorous Detector (NPD)

When this detector was invented by Karmen and Giuffrida in 1964 [26] it was known as the *alkali flame ionization detector (AFID)* because it consisted of an FID to which was added a bead of an alkali metal salt. As it has continued to evolve, its name has also changed and it has been known as a thermionic ionization detector (TID), a flame thermionic detector (FTD), a thermionic specific detector (TSD), etc.

Basically, Karmen and others have found that the FID shows selectively higher sensitivity when an alkali metal salt is present in the vicinity of the flame. In its present configuration, a bead of rubidium or cesium salt is electrically heated in the region where the flame ionization occurs. While

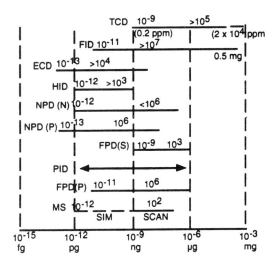

Fig. 7.15. Comparison of working ranges for common GC detectors.

the mechanism is not well understood, the detector does show enhanced detectivity for phosphorous-, nitrogen- and some halogen-containing substances.

Photoionization Detector (PID)

This ionization-type detector has also gone through several designs dating back to 1960. In its present form, an ultraviolet lamp (for example, 10.2 eV) emits sufficiently high energy photons to ionize directly many organic compounds. The resulting ions are collected and amplified to form the signal.

A related type of detector uses a spark to generate high energy photons that produce sample ionization. This detector is called a discharge ionization detector (DID). It finds application in the analysis of fixed gases at lower levels than can be determined with a TCD.

Flame Photometric Detector (FPD)

Flame photometry was adapted for use with an FID type flame for use in GC in 1966. The application to organic analysis is mainly for sulfur compounds (at 394 nm) and phosphorous compounds (at 526 nm) as found in pesticide residues and air pollutants.

Gas Density Balance (GADE)

This detector, which was invented by Martin and James [27] in 1956, is not widely used but is still commercially available and has some unique characteristics. It can be used for quantitative analysis without calibration if the densities of the analytes are known because that is the principle on which it is based. It can also be used to determine molecular weights of analytes if analyses are made in two different carrier gases [28].

Mass Selective Detector (MSD)

Mass spectrometers can be used as GC detectors. They need to have compatible characteristics and be properly coupled to the chromatograph. Some of them are referred to as mass selective detectors (MSD), which indicates that they are considered GC detectors, but the combined technique can also be called GC/MS, which indicates the coupling of two analytical instruments. Whatever the name, the use of a mass spectrometer with a gas chromatograph is a very powerful, useful, and popular combination, and it is treated in more detail in Chapter 10.

For more information on these and other detectors in Table 7.1, consult the references given in the table.

REFERENCES

1. David, D. J., *Gas Chromatographic Detectors,* Wiley, New York, 1974.
2. Hill, H. H., and McMinn, D. G., Eds., *Detectors for Capillary Chromatography,* Wiley, New York, 1992.
3. Scott, R. P. W., *Chromatographic Detectors: Design, Function and Operation,* Marcel Dekker, New York, 1996.
4. Sievers, R. E., Ed., *Selective Detectors: Enviromental, Industrial, and Biomedical Applications,* Wiley, New York, 1995.
5. Henrich, L. H., *Modern Practice of Gas Chromatography,* 3rd ed., R. L. Grob, Ed., Wiley, New York, 1995, Chapter 5.
6. Ouchi, G. I., *LC-GC* **14,** 472–476 (1996).
7. Johnson, E. L., and Stevenson, R., *Basic Liquid Chromatography,* Varian Associates, Palo Alto, CA, 1978, p. 278.
8. Ettre, L. S., *Pure Appl. Chem.,* **65,** 819–872 (1993).
9. MacDougall, D., et al., *Anal. Chem.,* **52,** 2242–2249 (1980).
10. *The United States Pharmacopeia, USP 23,* United States Pharmacopeial Convention, Inc., Rockville, MD, 1995.
11. Hinshaw, J. V., *LC-GC,* **14,** 950 (1996).
12. Sternberg, J. C., Gallaway, W. S., and Jones, D. T. C., *Gas Chromatography, Third International Symposium,* Instrument Society of America, Academic Press, 1962, pp. 231–267.
13. Sevcik, J., Kaiser, R. E., and Rieder, R., *J. Chromatogr.,* **126,** 361 (1976).
14. Holm, T., and Madsen, J. O., *Anal. Chem.,* **68,** 3607–3611 (1996).
15. Scanlon, J. T., and Willis, D. E., *J. Chromatogr. Sci.,* **23,** 333–340 (1985).
16. Simon, Jr., R. K., *J. Chromatogr. Sci.,* **23,** 313 (1985).
17. Wittebrood, R. T., *Chromatographia,* **5,** 454 (1972).
18. Pecsar, R. E., DeLew, R. B., and Iwao, K. R., *Anal. Chem.,* **45,** 2191 (1973).
19. Lochmuller, C. H., Gordon, B. M., Lawson, A. E., and Mathieu, R. J., *J. Chromatogr. Sci.,* **16,** 523 (1978).
20. Lawson, Jr., A. E., and Miller, J. M., *J. Chromatogr. Sci.,* **4,** 273 (1966).
21. Miller, J. M., and Lawson, Jr., A. E., *Anal. Chem.,* **37,** 1348–1351 (1965).
22. Purcell, J. E., and Ettre, L. S., *J. Chromatogr. Sci.,* **3,** 69 (1965).
23. Lovelock, J. E., *Anal. Chem.,* **33,** 162 (1961).
24. Grimsrud, E. P., *Detectors for Capillary Chromatography,* H. H. Hill and D. G. McMinn, Eds., Chpt. 5, Wiley, New York, 1992.
25. Mudabushi, J., Cai, H., Stearns, S., and Wentworth, W., *Am. Lab.,* **27,** (15), 21–30 (1995). Cai, H., Wentworth, W. E., and Stearns, S. D., *Anal. Chem.,* **68,** 1233 (1996).
26. Karmen, A., and Giuffrida, L., *Nature,* **201,** 1204 (1946).
27. Martin, A. J. P., and James, A. T., *Biochem. J.,* **63,** 138 (1956).
28. Liberti, A., Conti, L., and Crescenzi, V., *Nature,* **178,** 1067 (1956).

8. Qualitative and Quantitative Analysis

Gas chromatography can be used for both qualitative and quantitative analysis. Because it is more useful for quantitative analysis, most of this chapter is devoted to that topic; however, it begins with a brief look at qualitative analysis.

QUALITATIVE ANALYSIS

The chromatographic parameter used for qualitative analysis is the retention volume or some closely related parameter. However, since retention parameters cannot *confirm* peak identity, it is common to couple a mass spectrometer (MS) to the GC (GC–MS) for qualitative analysis. GC–MS is so widely used that it is discussed in detail in Chapter 10.

Table 8.1 lists the most common methods used for qualitative analysis in GC. Reference 1 is a good summary of these and other methods.

Retention Parameters

The retention time for a given solute can be used for its identity if the following column variables are kept constant: length, stationary phase and its thickness (liquid loading), temperature, and pressure (carrier gas flow

TABLE 8.1 GC Methods For Qualitative Analysis

Method	References
1. Retention parameters	
Retention time	1, 2
Relative retention time; Retention indexes	3, 4
2. Use of selective detectors	2
Dual channel GC	1, 5, 6
On-line	
MS or MSD (mass selective detector)	2, 5, 7
FTIR	2, 5, 8, 9, 10
Off-line	5
MS, MSD	
FTIR	
NMR	
UV	
3. Other methods	
Chemical derivatization	1, 2, 5
Pre-column	
Post-column	
Pyrolysis and chromatopyrography	11
Molecular weight chromatograph (gas density balance)	1, 2, 12

rate). As an example, consider that an unknown sample produced the chromatogram shown in Figure 8.1*a*. If one wished to know which of the components were *n*-alcohols, a series of *n*-alcohol standards could be run under identical conditions producing a chromatogram like Figure 8.1*b*. As shown in the figure, those peaks whose retention times exactly match those of the standards can be identified as the *n*-alcohols. This process will only work if the components of the unknown are alcohols.

This procedure will not be effective if the number of possible compounds is large—retention volumes are not that characteristic. Since there are over 30,000 organic compounds in common use, the gas chromatograph by itself cannot be used to identify a single compound from among this large group. Retention times are characteristic of a GC system, but they are not unique, so GC retention times cannot be used for qualitative confirmation.

However, *relative* retention volumes are more reproducible than individual retention volumes, so qualitative data should be reported on a relative basis. The retention index attributed to Kovats (see Chapter 4) has become a reliable method for reporting retention data. If a retention parameter is to be used for qualitative identification or classification, the Kovats retention index is a good method to choose.

Selective Detectors and Dual Detectors

Selective GC detectors can sometimes be used to help identify classes of compounds to which they show high sensitivity. The list of detectors in Chapter 7 can be consulted for further information and references.

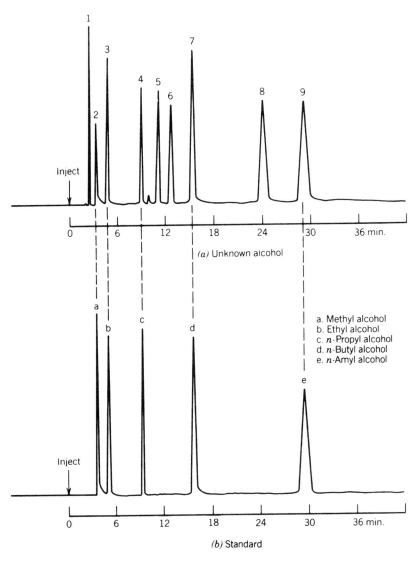

Fig. 8.1. Identification of unknown by retention times using standards. From Miller, J. M., *Chromatography: Concepts and Contrasts,* John Wiley & Sons, Inc., New York, 1987, p. 76. Reproduced courtesy of John Wiley & Sons, Inc.

More interesting is the use of two different detectors in parallel at the exit of a GC column—so called *dual channel* detection. The detectors chosen should have major differences in sensitivity for different classes of compounds. Both signals are recorded simultaneously producing parallel chromatograms like those shown in Figure 8.2. Identifications can be made

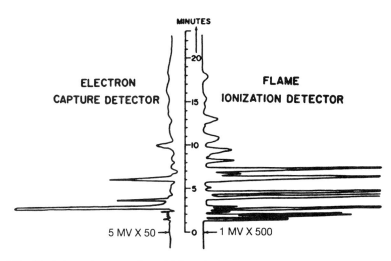

Fig. 8.2. Dual channel presentation of GC analysis of gasoline sample on a packed DC-200 column. Courtesy of Perkin-Elmer Corp. From Miller, J. M., *Chromatography: Concepts and Contrasts,* John Wiley & Sons, Inc., New York, 1987, p. 83. Reproduced courtesy of John Wiley & Sons, Inc.

by inspection of the chromatograms (Fig. 8.3) or from the ratios of the detector responses. The latter are often characteristic of classes of compounds. Figure 8.4 shows that the ratios from the data in Figure 8.3 clearly differentiate between paraffins, olefins, and aromatics in this example. When combined with the retention index, the ratio can lead to an identification of a particular homolog within a given class.

Off-line Instruments and Tests

In principle, one could collect the effluent from a GC column and identify it on any suitable instrument. A simple setup for collecting effluents in a cold trap is shown in Figure 8.5. The trapped sample could be transferred to an instrument for identification (MS, FTIR, NMR, UV), subjected to microanalysis, or reacted with a chemical reagent to produce a characteristic derivative. Commonly, the most useful instruments (MS and FTIR) are usually coupled on-line.

Other methods that can be used for identification are pyrolysis, derivatization, and the molecular weight chromatograph. References to these methods are given in Table 8.1.

On-line Instruments

GC–MS has already been mentioned as the premier method for qualitative analysis (see Chapter 10). A complementary identification technique is Fourier Transform infrared coupled to gas chromatography (GC–FTIR).

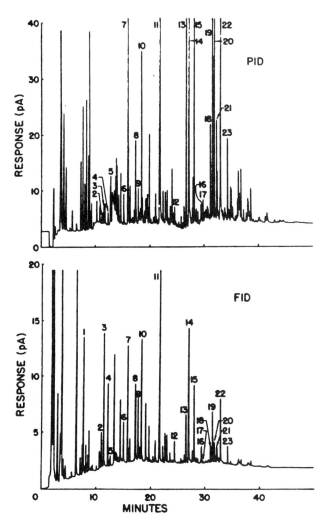

Fig. 8.3. Dual channel presentation of GC analysis of air contaminants in parking lot. Reprinted with permission from [13]. Copyright 1983, American Chemical Society. From Miller, J. M., *Chromatography: Concepts and Contrasts,* John Wiley & Sons, Inc., New York, 1987, p. 84. Reproduced courtesy of John Wiley & Sons, Inc.

The increased sensitivity of the Fourier Transform method of data handling has contributed greatly to its utility.

The two IR interfaces in common use are the light pipe [8] and so-called matrix isolation [9]. In the former method, the column effluent is passed through a heated IR gas cell (light pipe), and in the latter, it is condensed and frozen into a matrix suitable for analysis by IR [10].

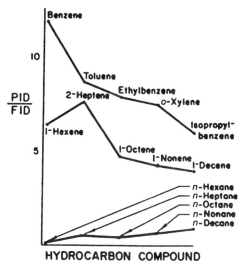

Fig. 8.4. Relative (PID/FID) response for 15 hydrocarbons. Reprinted with permission from [13]. Copyright 1983, American Chemical Society. From Miller, J. M., *Chromatography: Concepts and Contrasts,* John Wiley & Sons, Inc., New York, 1987, p. 86. Reproduced courtesy of John Wiley & Sons, Inc.

Since IR is nondestructive, it is possible to couple both the IR and the MS to the same gas chromatograph, producing GC–FTIR–MS. The special requirements and some applications have been described [8, 14].

QUANTITATIVE ANALYSIS

Making quantitative measurements is always accompanied by errors and necessitates an understanding of detectors (see Chapter 7) and data systems (see Chapter 2). Sampling, sample preparation, instrument and method validation, and quality assurance are all important parts of the process. Trace analysis, which is becoming increasingly popular, requires that all

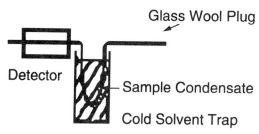

Fig. 8.5. Simple trapping device for qualitative analysis.

steps in the analysis be done with care. As an example of the guidelines that are common in trace analysis, the report of the American Chemical Society Subcommittee on Environmental Analytical Chemistry can be consulted [15]. It addresses the issues of data acquisition and data quality evaluation.

A short review of the statistical methods for handling error analysis is given here, followed by a brief discussion of typical errors. Then, the common methods of analysis are presented.

Statistics of Quantitative Calculations

Errors of measurement can be classified as *determinate* or *indeterminate*. The latter is random and can be treated statistically (Gaussian statistics); the former is not, and the source of the nonrandom error should be found and eliminated.

The distribution of random errors should follow the Gaussian or normal curve if the number of measurements is large enough. The shape of Gaussian distribution was given in Chapter 3 (Fig. 3.4). It can be characterized by two variables—the *central tendency* and the symmetrical *variation about the central tendency*. Two measures of the central tendency are the mean, $\overline{X}$, and the median. One of these values is usually taken as the "correct" value for an analysis, although statistically there is no "correct" value but rather the "most probable" value. The ability of an analyst to determine this most probable value is referred to as his *accuracy*.

The spread of the data around the mean is usually measured as the standard deviation, σ

$$\sigma = \sqrt{\frac{\Sigma (X - \overline{X})^2}{(n - 1)}} \tag{1}$$

where n is the number of measurements. The square of the standard deviation is called the *variance*. The ability of an analyst to acquire data with a small σ is referred to as his *precision*.

Precision and accuracy can be represented as shots at a target as shown in Figure 8.6. Figure 8.6*a* shows good accuracy and precision; Figure 8.6*b* good precision but poor accuracy; and Figure 8.6*c* poor precision that will result in poor accuracy unless a large number of shots are taken. The situation in Figure 8.6*b* suggests that a determinate error is present; maybe the gunsight is out of alignment.

Two other terms are in common use to distinguish two types of precision. One is *repeatability*, which refers to the precision in one lab, by one analyst, and on one instrument. The other is *reproducibility*, which refers to the precision among different labs and consequently different analysts and different instruments. We expect and usually find that reproducibility is not as good as repeatability.

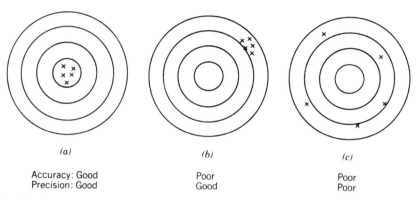

(a) (b) (c)

| Accuracy: Good | Poor | Poor |
| Precision: Good | Good | Poor |

Fig. 8.6. Illustrations of the definitions of accuracy and precision. From Miller, J. M., *Chromatography: Concepts and Contrasts,* John Wiley & Sons, Inc., New York, 1987, p. 99. Reproduced courtesy of John Wiley & Sons, Inc.

A related term used by the United States Pharmacopeia (USP) to specify instrument reproducibility is *ruggedness.* It expresses a rigorous test condition when the same test method is used in many different laboratories over an extended period of time.

In a set of data, a *relative standard deviation,* RSD, carries more information than the standard deviation itself. The relative standard deviation, or coefficient of variation as it is sometimes called is defined as:

$$RSD = \sigma_{rel} = \frac{\sigma}{\overline{X}} \tag{2}$$

The minimum information usually given to characterize the results of an analysis is one of each of the two variables we have discussed—usually the mean and the relative standard deviation. Table 8.2 contains two sets

TABLE 8.2   Comparison of the Precision of Two Analysts; Results of a GC Analysis of Methylethyl Ketone

| | Results Obtained by | |
	Chemist A	Chemist B
	10.0	10.2
	12.0	10.6
	9.0	9.8
	11.0	10.1
	8.0	9.3
Ave, $\overline{X}$	10.0	10.0
St. Dev., σ	1.58	0.48
RSD, σ_{rel}	15.8%	4.8%

of data obtained by two different analysts. While both have obtained the same average value, $\overline{X}$, chemist B has a smaller relative standard deviation and is therefore considered to be the better analyst or the one working with the better system.

One step in all quantitative procedures is the calibration step. Calibration is essential, and is often the limiting factor for obtaining accuracy in trace analysis. Good calibration and careful precision yield high accuracy.

Errors to be Avoided in Making Measurements

In a quantitative analysis, separation by gas chromatography is only one step in the total procedure. Errors that occur in any step can invalidate the best chromatographic analysis, so attention must be paid to all steps.

The steps in an analysis usually include the following: sampling, sample preparation and workup, separation (chromatography), detection of the analyte, data analysis including peak area integration, and calculations. With major advances in GC instrumentation and integration in the past 20 years, the major sources of GC error are usually sampling and sample preparation, especially if dirty matrices are involved.

In sampling, the objective is to get a small sample that is representative of the whole. Sample preparation can include such techniques as: grinding and crushing, dissolving, filtering, diluting, extracting, concentrating, and derivatizing. In each step care must be taken to avoid losses and contamination. If an internal standard (discussed later in this chapter) is used, it should be added to the sample before sample processing is begun.

The gas chromatographic separation should be carried out following the advice given in this and other chromatographic treatises; some objectives are: good resolution of all peaks, symmetrical peaks, low noise levels, short analysis times, sample sizes in the linear range of the detector, etc.

Data analysis and data systems are presented in Chapter 2. Of special interest is the conversion of the analog signal to digital data. This task can be accomplished by either of two ways—integration of the area under the peaks or measurement of the peak height. With today's electronic integrators and computers, peak area is the preferred method, especially if there may be changes in chromatographic conditions during the run, such as column temperature, flow rate, or sample injection reproducibility. However, peak height measurements are less affected by overlapping peaks, noise, and sloping baselines. In the discussions that follow, all data will be presented as peak areas.

Methods of Quantitative Analysis

Five methods of quantitative analysis will be discussed briefly, proceeding from the most simple and least accurate to those capable of higher accuracy.

Area Normalization

As the name implies, area normalization is really a calculation of area percent which is assumed to be equal to weight percent. If X is the unknown analyte,

$$\text{area } \%X = \left[\frac{A_x}{\sum\limits_i (A_i)}\right] 100 \tag{3}$$

where A_x is the area of X and the denominator is the sum of all the areas. For this method to be accurate, the following criteria must be met:

- All analytes must be eluted.
- All analytes must be detected.
- All analytes must have the same sensitivity (response/mass).

These three conditions are rarely met, but this method is simple and is often useful if a semiquantitative analysis is sufficient or if some analytes have not been identified or are not available in pure form (for use in preparing standards).

Area Normalization with Response Factors

If standards are available, the third limitation can be removed by running the standards to obtain relative response factors, f. One substance (it can be an analyte in the sample) is chosen as the standard, and its response factor f, is given an arbitrary value like 1.00. Mixtures, by weight, are made of the standard and the other analytes, and they are chromatographed. The areas of the two peaks—A_s and A_x for the standard and the unknown, respectively—are measured, and the relative response factor of the unknown, f_x, is calculated,

$$f_x = f_s \times \left(\frac{A_s}{A_x}\right) \times \left(\frac{w_x}{w_s}\right) \tag{4}$$

where w_x/w_s is the weight ratio of the unknown to the standard.

Relative response factors of some common compounds have been published for the most common GC detectors, and some representative values from an early work by Dietz [16] are given in Table 8.3 for the FID and TCD. These values are ± 3%, and since they were obtained using packed columns they may contain some column bleed. For the highest accuracy, one should determine his/her own factors.

When the unknown sample is run, each area is measured and multiplied by its factor. Then, the percentage is calculated as before:

TABLE 8.3 Relative Response Values for the FID and TCD (Wt. %) [16]*

Compound	Relative Response Factors, Wt. %	
	FID[a]	TCD
n-Paraffins		
Methane	1.03	0.45
Ethane	1.03	0.59
Propane	1.02	0.68
Butane	0.92	0.68
Pentane	0.96	0.69
Hexane	0.97	0.70
Octane	1.03	0.71
Branched Paraffins		
Isopentane	0.95	0.71
2,3-Dimethylpentane	1.01	0.74
2,2,4-Trimethylpentane	1.00	0.78
Unsaturates		
Ethylene	0.98	0.585
Aromatics		
Benzene	0.89	0.78
Toluene	0.93	0.79
o-Xylene	0.98	0.84
m-Xylene	0.96	0.81
p-Xylene	1.00	0.81
Oxygenated Compounds		
Acetone	2.04	0.68
Ethylmethylketone	1.64	0.74
Ethylacetate	2.53	0.79
Diethylether	—	0.67
Methanol	4.35	0.58
Ethanol	2.17	0.64
n-Propanol	1.67	0.60
i-Propanol	1.89	0.53
Nitrogen Compounds		
Aniline	1.33	0.82

* Reproduced from the *Journal of Chromatographic Science* by permission of Preston Publications, A Division of Preston Industries, Inc.

[a] FID response values are reciprocals of those given in the original publication by Dietz [16] so that they are consistent with the TCD values.

$$\text{Weight } \%X = \left[\frac{(A_x f_x)}{\sum_i (A_i f_i)}\right] 100 \qquad (5)$$

For example, consider a mixture of ethanol, hexane, benzene, and ethyl acetate being analyzed with a TCD. The areas obtained are given in Table 8.4 along with the response factors taken from Table 8.3. Each area is multiplied by its response factor:

TABLE 8.4   Example of Area Normalization With Response Factors

Compound	Raw Area	Wt. Response Factor	Corrected Area	Weight %	Area%	Abs. Error
Ethanol	5.0	0.64	3.20	17.6	20.0	+2.4
Hexane	9.0	0.70	6.30	34.7	36.0	+1.3
Benzene	4.0	0.78	3.12	17.2	16.0	−1.2
Ethylacetate	7.0	0.79	5.53	30.5	28.0	−2.5
Total	25.0	—	18.15	100.0	100.0	

$$\text{Ethanol: } (5.0) \times (0.64) = 3.20 \tag{6}$$

$$\text{Hexane: } (9.0) \times (0.70) = 6.30 \tag{7}$$

$$\text{Benzene: } (4.0) \times (0.78) = 3.12 \tag{8}$$

$$\text{Ethylacetate } (7.0) \times (0.79) = 5.53 \tag{9}$$

$$\text{Total} = 18.15 \tag{10}$$

Now, each corrected area is normalized to get the percent; for example:

$$\text{Ethanol: } \frac{3.20}{18.15} \times 100 = 17.6\% \tag{11}$$

This and the other values are given in Table 8.4 which contains the completed analysis (weight %) using response factors.

The errors that are incurred by not using response factors is also included in the last column of Table 8.4. They are the differences between the corrected weight % values and the (uncorrected) normalized area % values. For any given analysis the actual errors will depend upon the similarities or differences between the individual response values, of course. These calculations only serve as a typical example.

External Standard

This method is usually performed graphically and may be included in the software of the data system. Known amounts of the analyte of interest are chromatographed, the areas are measured, and a calibration curve is plotted. If the standard solutions vary in concentration, a constant volume must be introduced to the column for all samples and standards. Manual injection is usually unsatisfactory and limits the value of this method. Better results are obtained from autosamplers which inject at least one microliter.

If a calibration curve is not made and a data system is used to make the calculations, a slightly different procedure is followed. A calibration mixture prepared from pure standards is made by weight and chromatographed. An absolute calibration factor, equal to the grams per area produced, is stored in the data system for each analyte. When the unknown mixture is run, these factors are multiplied times the respective areas of each analyte in the unknown resulting in a value for the mass of each analyte. This procedure is a one-point calibration, as compared to the multipoint curve described before, and is somewhat less precise. Note also that these calibration factors are not the same as the relative response factors used in the area normalization method.

Internal Standard

This method and the next are particularly useful for techniques that are not too reproducible, and for situations where one does not (or cannot) recalibrate often. The internal standard method does not require exact or consistent sample volumes or response factors since the latter are built into the method; hence, it is good for manual injections. The standard chosen for this method can never be a component in a sample and it cannot overlap any sample peaks. A known amount of this standard is added to each sample—hence the name *internal* standard, I.S. The I.S. must meet several criteria:

- It should elute near the peaks of interest, but,
- It must be well resolved from them
- It should be chemically similar to the analytes of interest and not react with any sample components
- Like any standard, it must be available in high purity

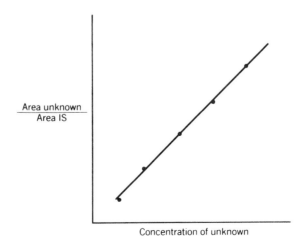

$\dfrac{\text{Area unknown}}{\text{Area IS}}$

Concentration of unknown

Fig. 8.7. Example of calibration plot using internal standard method. From Miller, J. M., *Chromatography: Concepts and Contrasts,* John Wiley & Sons, Inc., New York, 1987, p. 106. Reproduced courtesy of John Wiley & Sons, Inc.

The standard is added to the sample in about the same concentration as the analyte(s) of interest and prior to any chemical derivatization or other reactions. If many analytes are to be determined, several internal standards may be used to meet the preceding criteria.

Three or more calibration mixtures are made from pure samples of the analyte(s). A known amount of internal standard is added to each calibration mixture and to the unknown. Usually the same amount of standard is added volumetrically, e.g., 1.00 mL. All areas are measured and referenced to the area of the internal standard, either by the data system or by hand.

If multiple standards are used, a calibration graph like that shown in Figure 8.7 is plotted where both axes are relative to the standard. If the same amount of internal standard is added to each calibration mixture and unknown, the abscissa can simply represent concentration, not relative concentration. The unknown is determined from the calibration curve or from the calibration data in the data station. In either case, any variations in conditions from one run to the next are cancelled out by referencing all data to the internal standard. This method normally produces better accuracy, but it does require more steps and takes more time.

Some EPA methods refer to spiking with a standard referred to as a *surrogate*. The requirements of the surrogate and the reasons for using it are very similar to those of an internal standard. However, a surrogate is *not* used for *quantitative analysis* so the two terms are not the same and

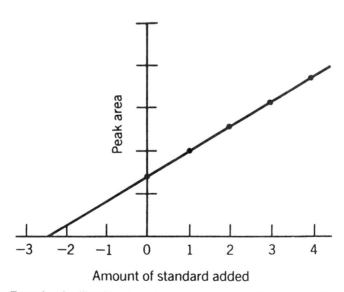

Fig. 8.8. Example of calibration plot using the standard addition method. From Miller, J. M., *Chromatography: Concepts and Contrasts,* John Wiley & Sons, Inc., New York, 1987, p. 107. Reproduced courtesy of John Wiley & Sons, Inc.

TABLE 8.5   Example of GC Quantitative Analysis

Component	True Weight (%)	Determined by GC (%) ± SD	Relative Error (%)
n-C_{10}	11.66	11.54 ± 0.02	1.0
n-C_{11}	16.94	16.91 ± 0.02	0.2
n-C_{12}	33.14	33.17 ± 0.02	0.1
n-C_{13}	38.26	38.38 ± 0.03	0.3

should not be confused with each other. In general, *spiking* standards are used to evaluate losses and recoveries during sample workup.

Standard Addition

In this method the standard is also added to the sample, but the chemical chosen as the standard is the *same* as the analyte of interest. It requires a highly reproducible sample volume, a limitation with manual syringe injection.

The principle of this method is that the additional, incremental signal produced by adding the standard is proportional to the amount of standard added, and this proportionality can be used to determine the concentration of analyte in the original sample. Equations can be used to make the necessary calculations, but the principle is more easily seen graphically. Figure 8.8 shows a typical standard addition calibration plot. Note that a signal is present when no standard is added; it represents the original concentration, which is to be determined. As increasing amounts of standard are added to the sample, the signal increases, producing a straight line calibration. To find the original "unknown" amount, the straight line is extrapolated until it crosses the abscissa; the absolute value on the abscissa is the original concentration. In actual practice, the preparation of samples and the calculation of results can be performed in several different ways [17].

Matisova and co-workers [18] have suggested that the need for a reproducible sample volume can be eliminated by combining the standard addition method with an *in situ* internal standard method. In the quantitative analysis of hydrocarbons in petroleum, they chose ethyl benzene as the standard for addition, but they used an unknown neighboring peak as an internal standard to which they referenced their data. This procedure eliminated the dependency on sample size and provided better quantitation than the area normalization method they were using.

Summary

GC results can be very accurate, down to about 0.1% RSD in the ideal case. Some typical results are shown in Table 8.5.

REFERENCES

1. Debbrecht, F. J. *Modern Practice of Gas Chromatography,* 3rd ed. R. L. Grob, Ed., Wiley, New York, 1995, pp. 393–425.
2. Leathard, D. A., *Advances in Chromatography,* Vol. 13, J. C. Giddings, Ed., Dekker, NY, 1975, pp. 265–304.
3. Blomberg, L. G., *Advances in Chromatography,* Vol. 26, J. C. Giddings, Ed., Dekker, NY, 1987, Chpt. 7, pp. 277–320.
4. Haken, J. K., *Advances in Chromatography,* Vol. 14, J. C. Giddings, Ed., Dekker, NY, 1976, Chpt. 8.
5. Ettre, L. S., and McFadden, W. H., Eds., *Ancillary Techniques of Gas Chromatography,* Wiley-Interscience, New York, 1969.
6. Krull, I. S., Swartz, M. E., and Driscoll, J. N. *Advances in Chromatography,* Vol. 24, J. C. Giddings, Ed., Dekker, NY, 1984, Chpt. 8, pp. 247–316.
7. Masucci, J. A., and Caldwell, G. W., *Modern Practice of Gas Chromatography,* Third Edition, R. L. Grob, Ed., Wiley, New York, 1995, pp. 393–425.
8. Leibrand, R. J., Ed., *Basics of GC/IRD and GC/IRD/MS,* Hewlett-Packard, Wilmington, DE, 1993.
9. Coleman III, W. M., and Gordon, B. M., *Advances in Chromatography,* Vol. 34, Brown, P. R., and Grushka, E., Eds., Dekker, NY, 1994, Chpt. 2, pp. 57–108.
10. Schreider, J. F., Demirian, J. C., and Stickler, J. C., *J. Chromatogr. Sci.,* **24,** 330 (1986).
11. Hu, J. C., *Advances in Chromatography,* Vol. 23, J. C. Giddings, Ed., Dekker, NY, 1984, Chpt. 5, pp. 149–198.
12. Bennet, C. E., DiCave, Jr., L. W., Paul, D. G., Wegener, J. A., and Levase, L. J., *Am. Lab.,* **3,** (5), 67 (1971).
13. Nutmagul, W., Cronn, D. R., and Hill, Jr., H. H., *Anal. Chem.,* **55,** 2160 (1983).
14. Wilkins, C. L., *Science* **222,** 291 (1983).
15. Macdougall, D., et al., *Anal. Chem.,* **55,** 2242–2249 (1980).
16. Dietz, W. A., *J. Chromatogr. Sci.,* **5,** 68 (1967).
17. Bader, M., *J. Chem. Educ.,* **57,** 703 (1980).
18. Matisova, E., Krupcik, J., Cellar, P., and Garaj, J., *J. Chromatogr.,* **303,** 151 (1984).

9. Programmed Temperature

Programmed temperature gas chromatography (PTGC) is the process of increasing the column temperature during a GC run. It is very effective method for optimizing an analysis and is often used for screening new samples. Before describing it in detail, let us consider the general effects of temperature on gas chromatographic results.

TEMPERATURE EFFECTS

Temperature is one of the two most important variables in GC, the other being the nature of the stationary phase. Retention times and retention factors decrease as temperature increases because the distribution constants are temperature-dependent in accordance with the Clausius–Clapeyron equation,

$$\log p^0 = - \frac{\Delta \mathcal{H}}{2.3 \, \mathcal{R}T} + \text{constant} \qquad (1)$$

where $\Delta \mathcal{H}$ is the enthalpy of vaporization at absolute temperature, T; $\mathcal{R}$ is the gas constant; and p^0 is the compound's vapor pressure at this temperature. The equation indicates that as the (absolute) temperature decreases the vapor pressure of the solute decreases logarithmically. A decrease in vapor pressure results in a decrease in the relative amount of solute in the mobile phase, viz, an increase in the retention factor, k, and an increase in retention time.

Figure 9.1, a plot of the log of net retention volume versus $1/T$ for a few typical solutes illustrates this relationship. Straight lines are obtained over a limited temperature range in accordance with our prediction based on equation 1. The slope of each line is proportional to that solute's enthalpy of vaporization and can be assumed to be constant over the temperature range shown.

To a first approximation, the lines in Figure 9.1 are parallel, indicating that the enthalpies of vaporization for these compounds are nearly the same. A closer inspection reveals that many pairs of lines diverge slightly

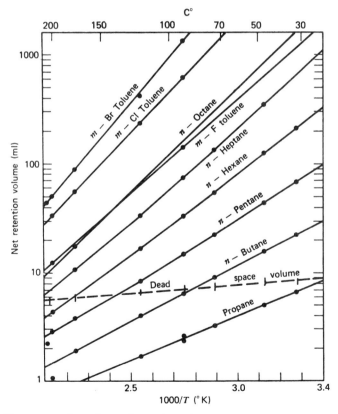

Fig. 9.1. Temperature dependence of retention volume. Reprinted with permission from [2]. Copyright 1964, Pergamon Journals, Ltd. From Miller, J. M., *Chromatography: Concepts and Contrasts,* John Wiley & Sons, Inc., New York, 1987, p. 145. Reproduced courtesy of John Wiley & Sons, Inc.

at low temperatures. From this observation we can draw the useful generalization that *GC separations are usually better at lower temperatures.* But look at the two solutes, *n*-octane and *m*-fluorotoluene; their lines cross at about 140°C. At that temperature, 140°C, they cannot be separated; at a lower temperature the toluene elutes first, but at a higher temperature the *reverse* is true. While it is not common for elution orders to reverse, it can happen, resulting in misidentification of peaks. See for example the work of Hinshaw with chlorinated pesticides [1].

Below is a list of the consequences of increasing the temperature for a GC analysis.

Effect of Temperature Increase

* Retention time and retention volume decrease
* Retention factor decreases
* Selectivity (α) changes (usually decreases)
* Efficiency (N) increases slightly

The effect of temperature on efficiency is quite complex [2] and does not always increase. Usually it is a minor effect and less important than the effect on column thermodynamics (selectivity). Overall, however, temperature effects are very significant and PTGC is very powerful.

ADVANTAGES AND DISADVANTAGES OF PTGC

If a sample being analyzed by GC contains components whose vapor pressures (boiling points) extend over a wide range, it is often impossible to select *one* temperature which will be suitable for an isothermal run. As an example, consider the separation of a wide range of homologs like the kerosene sample shown in Figure 9.2*a*. An isothermal run at 150°C prevents the lighter components ($<C_8$) from being totally separated and still takes over 90 minutes to elute the C_{15} paraffin which looks like the last one. Even so, this is probably the best isothermal temperature for this separation.

The separation can be significantly improved using programmed temperature. Figure 9.2*b* shows one such programmed run in which the temperature starts 50°C, less than the isothermal temperature used in Figure 9.2*a,* and is programmed at 8 degrees per minute up to 250°C, a temperature higher than the isothermal temperature. Increasing the temperature during the run decreases the partition coefficients of the analytes still on the column, so they move faster through the column, yielding decreased retention times.

Some major differences between the two runs illustrate the advantages of PTGC. For a homologous series, the retention times are logarithmic under isothermal conditions, but they are linear when programmed. The programmed run facilitated the separation of the low-boiling paraffins,

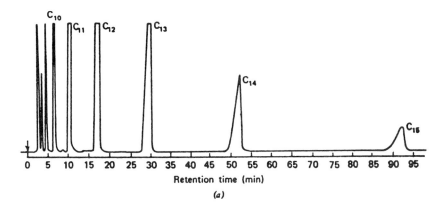

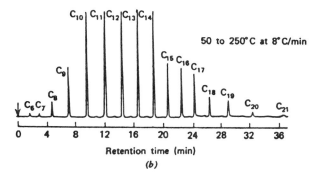

Fig. 9.2. Comparison of (*a*) isothermal and (*b*) programmed temperature separations of *n*-paraffins. From Miller, J. M., *Chromatography: Concepts and Contrasts,* John Wiley & Sons, Inc., New York, 1987, p. 146. Reproduced courtesy of John Wiley & Sons, Inc.

easily resolving several peaks before the C_8 peak while increasing the number of paraffins detected. The C_{15} peak elutes much faster (in about 21 minutes) and it turns out that it is not the last peak—six more hydrocarbons are observed by PTGC. All of the peak widths are about equal in PTGC; in the isothermal run, some fronting is evidenced in the higher boilers. Since the peak widths do not increase in PTGC, the heights of the late-eluting analytes are increased (peak areas are constant), providing better detectivity. The list below summarizes the advantages and disadvantages of PTGC.

Advantages and Disadvantages of PTGC

Advantages

1. Good scouting tool (rapid)
2. Shorter analysis times for complex samples

3. Better separation of wide boiling point range
4. Improved detection limits, peak shapes, and precision, especially for late eluting peaks
5. Excellent means of column cleaning

Disadvantages

1. More complex instrumentation required
2. Noiser signals at high temperatures
3. Limited number of suitable liquid phases
4. May be slower, considering cooling time

Another example of the application of PTGC for optimizing a separation is shown in Figure 9.3. Here multiple programming steps are used to get the optimum separation in the minimum time. Modern programmers typically provide as many as five temperature ramps.

Programmed temperature operation is good for screening new samples. A maximum amount of information about the sample composition is obtained in minimum time. Usually one can tell when the entire sample has been eluted, often a difficult judgment to make with isothermal operation.

REQUIREMENTS OF PTGC

PTGC requires a more versatile instrument than does isothermal GC. The major requirements are listed on the next page.

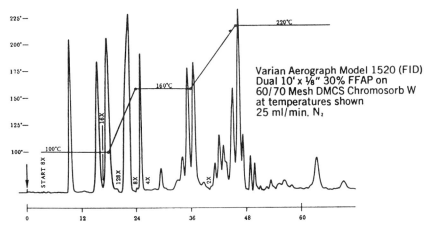

Varian Aerograph Model 1520 (FID)
Dual 10' x ⅛" 30% FFAP on
60/70 Mesh DMCS Chromosorb W
at temperatures shown
25 ml/min. N_2

Fig. 9.3. Multistep PTGC of orange oil.

Instrumentation Requirements for PTGC

1. Dry carrier gas
2. Temperature programmer
3. Three separate ovens (injector, column, detector)
4. Flow controller (differential pneumatic or electronic)
5. Dual column or column compensation to remove drift
6. Suitable liquid phase

Most important is the ability to control the programmed temperature increase in the column oven while keeping the detector and injection port at constant temperatures. An electronic temperature programmer is needed along with an oven design that has a low mass, a high volume fan, and a vent to outside air, also controlled by the programmer.

Some means is usually provided to control the carrier gas flow. In a packed column chromatograph, this is usually accomplished with a differential pneumatic flow control valve placed in the gas line upstream of the injection port.

In a capillary column chromatograph, constant *pressure* regulation is required for split/splitless sampling and a flow control valve cannot be used. As a consequence, the flow rate of carrier gas decreases during the programmed temperature run due to the increase in gas viscosity. Since the pressure drop across an OT column is relatively low, the change in flow rate is less severe than in packed columns. One solution is to set the initial flow rate above the optimum value and closer to the flow expected about 70% of the way through the program. This will insure an adequate flow at the higher temperatures. However, electronic pressure control (EPC) is available on some instruments and it can be used to maintain a constant flow by increasing the pressure during the run [3].

Other requirements are placed on the carrier gas and the stationary phase. As indicated in the PTGC instrumentation list, the carrier gas must be dry to prevent the accumulation of water (and other volatile impurities) at the cool column inlet (before the start of a run) since this phenomenon will result in ghost peaks during the PTGC run. One common solution to this problem is to insert a 5Å molecular sieve dryer in the gas line before the instrument.

Below is a list of some of the requirements of liquid phases.

Requirements of Liquid Phases for PTGC

1. Wide temperature range (200°C) with low vapor pressure over entire range
2. Reasonable viscosity at low temperature (for high N)
3. Selective solubility (for high α)

TABLE 9.1   **High Temperature Liquid Phases for PTGC**

	Liquid Phases	Temperature Range (°C)
Packed Column		
Nonpolar	OV-1®, SE-30®	100 to 350
	Dexsil 300®	100 to 400
Polar	Carbowax 20M®	60 to 225
	OV-17®	0 to 300
	OV-275®	25 to 250
Capillary Column		
Nonpolar	DB-1®	−60 to 360
	DB-5®	−60 to 360
Polar	DB-1701®	−20 to 300
	DB-210®	45 to 260
	DB Wax®	20 to 250

Liquid phases that meet these requirements and have been found useful are included in Table 9.1. Recall that fused silica OT columns that are coated with polyimide cannot be used above 380°C without degrading the coating.

THEORY OF PTGC

The theory of PTGC has been thoroughly treated by Harris and Habgood [4] and by Mikkelsen [5]. The following discussion has been taken from a simple but adequate treatment by Giddings [6].

The dependence of retention volume on temperature was illustrated in Figure 9.1. Let us determine what temperature increase is necessary to reduce the adjusted retention volume by 50%; that is:

$$\frac{(V_R')_1}{(V_R')_2} = \frac{1}{2} \tag{2}$$

Since the ratio of adjusted retention volumes is inversely proportional to the log of the ratio of solute vapor pressures according to the integrated form of the Clausius-Clapeyron equation, we can conclude that,

$$ln\frac{P_2}{P_1} = ln\ 2 = \frac{\Delta\mathcal{H}\Delta T}{\mathcal{R}T_1 T_2} \tag{3}$$

where ΔT is the difference between the two temperatures T_1 and T_2. Taking the logarithm and rearranging it, we get,

$$\Delta T = \frac{0.693\ \mathcal{R}T^2}{\Delta\mathcal{H}} \tag{4}$$

Assuming Trouton's Rule that $\Delta \mathcal{H}/T_{boil} = 23$ and a boiling temperature of 227°C (500°K) for a typical sample,

$$\Delta T = \frac{(0.693)\ (2)\ (500)^2}{23\ (500)} \approx 30°C \qquad (5)$$

As an approximation, then, an increase in temperature of 30°C will cut the retention volume in half. This rule of thumb is also useful for isothermal operations.

The effect of temperature programming on the migration of a typical analyte through a column is shown in Figure 9.4, where the 30-degree value is used to generate the step function. Hence, the relative migration rate will double for every 30°C. Final elution from the column is arbitrarily taken as occurring at 265°C, as shown in the figure. In actuality, the movement of the analyte through the column could proceed by the smooth curve (also

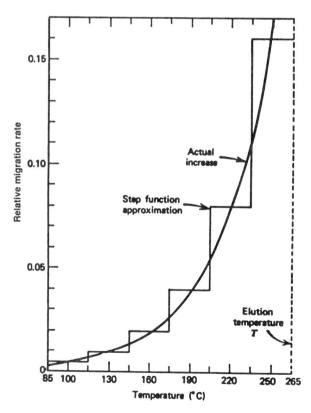

Fig. 9.4. Step-function approximation to programmed temperature GC. Reprinted with permission from the *Journal of Chemical Education*, Vol. 39, No. 11, November 1962, p. 571, copyright© 1962, Division of Chemical Education, Inc.

shown in the figure) since the temperature programming would be gradual and not stepwise, as assumed by our model.

If x is taken as the distance the analyte moved through the column in the last 30-degree increment, then one-half x is the distance it moved in the previous 30 degrees, one-quarter x in the 30 degrees before that, and so on. The sum of these fractions approaches 2, which must equal the total length L of the column ($2x = L$). Hence, the analyte traveled through the last half of the column in the last 30°C, through 3/4 of the column in 60°C, etc. Initially the solute was "frozen" at the inlet of the column, but when it began to migrate, its rate of migration doubled for every 30-degree increase in temperature.

The operation of PTGC can be envisioned as follows: the sample is injected onto the end of the cool column, and its components remain condensed there; as the temperature increases, the analytes vaporize and move down the column at increasing rates until they elute. It is for these reasons that the injection technique is not critical in PTGC and that all peaks have about the same peak widths—they spend about the same amount of time actively partitioning down the column.

For a variety of reasons, isothermal operation is often preferred in the workplace. If an initial screening is done by PTGC, one might wish to know which isothermal temperature would be the best one to use. Giddings has called this isothermal temperature the *significant temperature, T'*. Using reasoning based on the 30-degree value, he has found that

$$T' = T_f - 45 \tag{6}$$

where T_f is the final temperature, the temperature at which the analyte(s) of interest eluted in the PTGC run. Thus, for example, a solute eluting at a temperature of 225°C on a PTGC run would be best run isothermally at 180°C.

Three other important variables are the programming rate, the flow rate, and the column length. In general, one does not vary the length but uses a short column (and lower temperatures) and relatively high flow rates. The programming rate is often chosen to be fast enough to save time but slow enough to get adequate separations, somewhere between 4 and 10°C/min. However, for OT columns, one group of workers concluded that slow rates (around 2.5°C/min) and high flow rates (about 1 mL/min) are preferable [7]. Another study by Hinshaw [1] of a chlorinated pesticide mixture found that 8°C/min was preferable to either slower (down to 1.5°C/min) or faster (up to 30°C/min) rates.

SPECIAL TOPICS

In this section some topics related to temperature programming will be briefly discussed.

Quantitative Analysis

The data presented in this chapter clearly show the effect of PTGC on the size and shape of the individual peaks. This might lead one to conclude that PTGC cannot be used for quantitative analysis. This is not the case.

Consider the data given in Table 9.2 for the analysis of a synthetic mixture of *n*-paraffins analyzed by PTGC and isothermal GC. They show no significant difference between PTGC and isothermal GC when calibrations are carried out consistently by either technique. Modern instruments have the ability to maintain a constant temperature on the detector even during programmed temperature operation of the column, so that quantitation by the detector is unaffected and independent of column temperature.

Cryogenic Operation

Some chromatographs are provided with ovens that can be operated below ambient temperature, thus extending the range of temperature programming. Examples can be found in the extensive review of cryogenic GC by Brettell and Grob [8].

High Temperature GC

There has always been an interest in pushing GC to the highest temperatures possible. Several commercial instruments have upper temperature limits on column ovens and detector ovens of 400°C. Few columns can be operated at that high a temperature, but work has been reported in which the columns are routinely programmed up to 400°C. This research has given rise to a special technique called *high-temperature GC,* HTGC, defined as routine column temperature in excess of 325°C [9].

In a HTGC experiment, a short, lightly loaded OT column is employed, in accordance with the existing knowledge of GC theory. The most difficult aspect of HTGC, and the one to which most attention has been focused, is sample introduction. Conventional split/splitless methods for OT columns are not suitable because of the discrimination of high-boiling components that occurs in the vaporization process. On-column techniques do work but lead to considerable contamination of the column inlet. Programmed

TABLE 9.2 Comparison of Typical Quantitative Data

	Weight Percentage		
Sample	Actual	Isothermal	PTGC
Decane	11.66	11.54	11.66
Undecane	16.94	16.91	17.07
Dodecane	33.14	33.17	33.17
Tridecane	38.26	38.38	38.12

temperature *injection* (PTV) which works nicely with normal-temperature GC has been shown to be ideal for HTGC [9].

By programming the injector as high as 600°C, high molecular weight samples have been run successfully. One recent paper [9] reports the analysis of a polyethylene standard with an average molecular weight of 1000 Daltons and the successful separation and detection of the 100-carbon polymer with a molecular weight around 1500. More details can be found in this paper and the references listed in it.

REFERENCES

1. Hinshaw, J. V., *LC-GC*, **9,** 470 (1991).
2. Harris, W. E., and Habgood, H. W., *Talanta*, **11,** 115 (1964).
3. Stafford, S. S., Ed., *Electronic Pressure Control in Gas Chromatography*, second printing, Hewlett-Packard Co., Wilmington, DE, 1994. Hewlett-Packard also has available EPC software called "HP Pressure/Flow Calculator."
3. Harris, W. E., and Habgood, H. W., *Programmed Temperature Gas Chromatography*, Wiley, New York, 1966.
5. Mikkelsen, L., *Adv. Chromatogr, N. Y.*, **2,** 337 (1966).
6. Giddings, J. C., *J. Chem. Educ.* **39,** 569 (1962).
7. Jones, L. A., Kirby, S. L., Garganta, C. L., Gerig, T. M., and Mulik, J. D., *Anal. Chem.* **55,** 1354 (1983).
8. Brettell, T. A., and Grob, R. L., *Am. Lab.*, **17**(10), 19 and (11), 50 (1985).
9. van Lieshout, H. P. M., Janssen, H. G., and Cramers, C. A., *Am. Lab.*, **27,** (12), 38 (1995).

10.   Special Topics

Even in a basic textbook like this one there is a need to discuss special topics like GC–MS, chiral separations, sample preparation, and derivatives. Much current research is focused on these special techniques coupled to GC. The most important one is GC–MS, the common acronym for the technique in which a gas chromatograph is directly coupled to a bench-top mass spectrometer. The mass spectrometer was mentioned in Chapter 7 as one special GC detector, the MSD, but it is covered more thoroughly in this chapter.

Other topics that are covered briefly are: the separation of chiral compounds, some special sampling techniques (headspace and solid phase microextraction) and derivatization.

GC–MS

Today it is estimated that there are over 25,000 bench-top GC–MS systems worldwide and annual sales exceed 2,000 units per year. What makes this combination so powerful and so popular?

As we have noted earlier, GC is the premier analytical tehnique for the separation of volatile compounds. It combines speed of analysis, resolution, ease of operation, excellent quantitative results, and moderate costs. Unfortunately, GC systems cannot confirm the identity or structure of any peak. Retention times are related to partition coefficients (Chapter 3), and while they are characteristic of a well-defined system, they are not unique. GC data alone cannot be used to identify peaks.

Mass spectroscopy on the other hand is one of the most information-rich detectors. It requires only micrograms of sample, but it provides data for both the qualitative identification of unknown compounds (structure, elemental composition, and molecular weight), as well as their quantitation. In addition, it is easily coupled to a GC system.

More complete information about GC–MS can be found in the monographs listed [1–5], but an introduction to the topic follows.

Instrumentation

Figure 10.1 is a schematic of a typical low resolution mass spectrometer of the type commonly used with GC. Because of its small size, it is often referred to as a *bench-top* MS.

Sample Inlets

A sample inlet allows for the introduction of a very small amount of sample from a variety of sources. A large gas bulb can be used to introduce gaseous samples through a small pinhole into the ionization source. An inlet with septum would allow easy introduction of liquids, or solutions of solids; and finally, a vacuum interlock system is a common means for the introduction

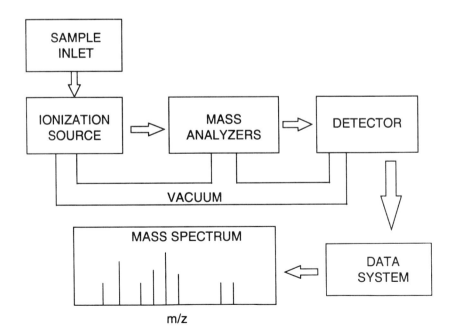

Fig. 10.1. Schematic of a mass spectrometer.

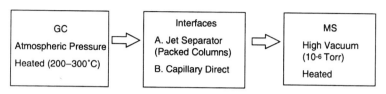

Fig. 10.2.   Coupling of GC to MS.

of solids. For use with a chromatograph, a variety of other methods have
been used.

Figure 10.2 shows schematically the coupling of a GC system to an MS
system. Both systems are heated (200–300°C), both deal with compounds
in the vapor state, and both require small samples (micro- or nano-grams).
GC and MS systems are very compatible. The only problem is that the
atmospheric pressure output of the GC must be reduced to a vacuum of
10^{-5} to 10^{-6} torr for the MS inlet. The coupling of the two must be done
with a reduction of pressure, and is accomplished with an interface.

Figure 10.3A shows a common interface in use today. Most GC–MS
systems use capillary columns, and fused silica tubing permits an inert, high
efficiency direct transfer between the two systems. For capillary flow rates
of 5 mL/min or less, a direct interface is possible. Bench-top GC–MS
systems can easily handle these low flow rates, and they provide better
sensitivity (transfer of total sample) and better preservation of GC results.

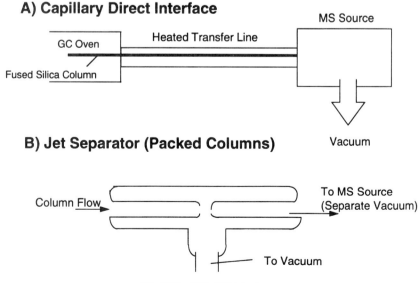

Fig. 10.3.   GC–MS interfaces.

Older GC–MS systems used packed columns, usually 2 mm i.d., with flow rates of about 30 mL/min. These packed column systems required an interface like the jet separator shown in Figure 10.3B. This separator consists of 2 glass tubes aligned with a small distance (~ 1mm) between them. Most of the carrier gas (usually He) entering from the GC column is pumped away by a separate vacuum system. The larger sample molecules maintain their momentum and pass preferentially into the second capillary and into the MS source. Sample enrichment occurs and the initial atmospheric pressure is greatly reduced, allowing the MS vacuum to handle the smaller flow rate. Both temperature and surface activity of the glass jet separator must be carefully controlled both to maximize sample transfer and to preserve sample integrity.

Ionization Sources

Analyte molecules must first be ionized in order to be attracted (or repelled) by the proper magnetic or electrical fields. There are numerous ionization techniques, but electron impact (EI) is the oldest, most common and most simple. The ionization source is heated and under vacuum so most samples are easily vaporized and then ionized. Ionization is usually accomplished by impact of a highly energetic (70 ev) electron beam.

A typical source is shown schematically in Figure 10.4. Effluent from the GC column passes into a heated ionization source at low vacuum. Electrons are drawn out from a tungsten filament by a collector voltage of 70 ev. The voltage applied to the filament defines the energy of the electrons. These high energy electrons strike the neutral analyte molecules, causing

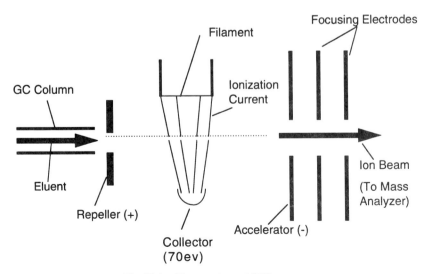

Fig. 10.4.   Electron impact (EI) source.

ionization (usually loss of an electron) and fragmentation. This ionization technique produces almost exclusively positive ions:

$$M + e^- \rightarrow M^+ + 2e^- \tag{1}$$

Alternate means of achieving ionization include chemical ionization (CI), negative chemical ionization (NCI), and fast atom bombardment (FAB). In CI, a reagent gas like methane is admitted to the ion chamber where it is ionized, producing a cation that undergoes further reactions to produce secondary ions. For example:

$$CH_4 + e^- \rightarrow CH_4^+ + 2\ e^- \tag{2}$$

$$CH_4^+ + CH_4 \rightarrow CH_5^+ + CH_3 \tag{3}$$

The secondary ion (CH_5^+ in this example) serves as a reagent to ionize the sample gently. Usually this process results in less fragmentation and more simple mass spectra. The major MS peaks that normally result are ($M + 1$), (M), ($M - 1$), and ($M + 29$), where M is the mass of the analyte being studied.

To perform chemical ionization, the ion volume of the spectrometer is usually different from the one used for EI, the operating pressure is higher (partially due to the additional reagent gas), and the temperature is lower. Certain types of molecules also yield good negative ion spectra by NCI, providing another option for analysis. However, most bench-top GC–MS instruments are not capable of CI operation.

A comparison of CI and EI spectra is shown in Figure 10.5 for ortal, a barbiturate with a molecular weight of 240. The base peak in the CI spectrum is 241, the expected ($M + 1$) peak. There are some other, small peaks, but this spectrum shows the value of the CI method in providing an assignment of the molecular weight. The EI spectrum, on the other hand, shows a very small parent ion with major peaks at 140 and 156. These fragment ions can be used to aid in the assignment of the structure—information not provided by CI.

Analyzers and Detectors

After ionization, the charged particles are repelled and attracted by charged lenses into the mass analyzer. Here the ionic species are separated by their mass-to-charge ratio (m/z) by either magnetic or electrical fields. Typical mass analyzers for GC–MS are quadrupoles or ion traps. Other analyzers are: single-focusing magnetic sector; double-focusing magnetic sector (high resolution, more expensive); and time of flight.

The quadrupole mass analyzer consists of four hyperbolic rods at right angles to each other (see Fig. 10.6). A DC voltage is applied to all rods (adjacent rods have opposite signs) and the signs of the voltage are rapidly

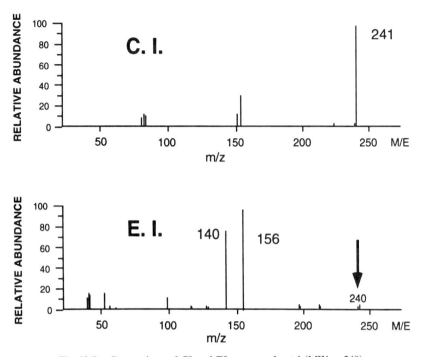

Fig. 10.5. Comparison of CI and EI spectra of ortal (MW = 240).

reversed. A radio frequency is also applied to the four rods. Depending on the combination of the radio frequency and the direct current potentials, ions of only one mass-to-charge ratio will pass through the rods and reach the detector. Ions with other m/z ratios will strike the rods and be annihilated.

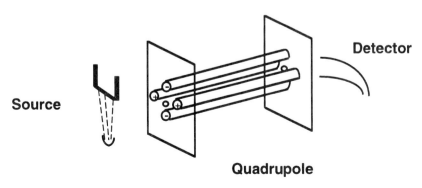

Fig. 10.6. Quadrupole mass analyzer.

The quadrupole analyzer has the advantages of simplicity, small size, moderate cost, and rapid scanning which make it ideal for GC–MS systems. It is restricted to about 2,000 Daltons and has low resolution when compared to double focusing mass spectrometers.

Figure 10.7 shows schematically an ion trap mass analyzer which was developed specifically for GC–MS. It is a simpler version of the quadrupole in which the ring electrode, having only a radio frequency applied to it, serves essentially as a monopole to define a stable region for charged species inside the circular electrode space. There are two end caps on the top and bottom of the circular ring electrode. Effluent from the GC enters the top end cap; some analytes are ionized and then trapped in stable trajectories inside the ring electrode. The radio frequency can be altered to eject sequentially ions with selected m/z ratios from the ion trap and pass them through the end cap to the detector.

Ion traps are also simple in design, modest in cost, and capable of rapid scanning for GC–MS applications. The spectra generated often differ from classical quadrupole spectra, and some ions may undergo dissociation or ion/molecule collisions inside the ion trap.

After separation of the ions produced, a detector, usually a continuous dynode version of an electron multiplier, is used to count the ions and generate a mass spectrum. Such a detector is shown schematically in Figure 10.8. Ions from the mass analyzer strike the semi-conductive surface and release a cascade of electrons. These are accelerated by a potential difference to another portion of the semi-conductive surface where a larger cascade of electrons results. This process is repeated several times until amplification of the original weak input is magnified about 1-million-fold.

Note that the entire MS system is under high vacuum. This is an essential requirement to avoid the loss of the charged species by collision with other ions, molecules, or surfaces.

The *mass spectrum* is simply a plot of the ion abundance as a function of m/z. Under controlled conditions, the ratios of ion abundance and the

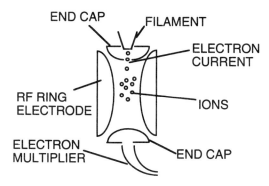

Fig. 10.7. Ion trap mass analyzer.

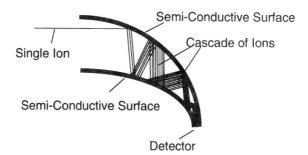

Fig. 10.8. Electron multiplier (continuous-dynode version).

specific m/z species present are unique for each compound. They can be used to establish the molecular weight and the chemical structure of each compound.

History

After J. J. Thompson used a mass spectrometer to separate atomic isotopes in 1913, MS was slowly developed and improved as an analytical tool. It proved to be powerful for identifying unknown compounds, as well as elucidating structures of both inorganic and organic compounds. It was widely used for the characterization of petroleum products and probably would have grown even more dramatically if GC had not been introduced in 1952.

MS was first coupled to GC in 1959 by Gohlke [6]. The early instruments were expensive, cumbersome and complex, usually requiring considerable expertise and maintenance to keep them running. Magnetic sector instruments were the most popular type of analyzer; however, they could not provide the rapid scanning (few seconds) necessary to generate a mass spectrum of a peak eluting from a GC. Faster scanning analyzers would be required.

By the late 1960s, it was obvious that GC was a huge analytical market and growing rapidly, but no GC detector provided as much information as was available from MS. Thus was born the dedicated "GC–MS" systems where a new MS was designed as a detector for the GC.

By specifying that the sample inlet would be a GC, the MS requirements could be simplified. The mass range could be limited to about 600 Daltons; low resolution was satisfactory because the GC possessed high resolution capabilities so the eluting peaks would in most cases be "pure". The tough part was the development of rapid scanning devices (hopefully 40–400 Daltons several times per second) and more simple, rugged instruments which could be used in routine analytical laboratories. The quadrupole, and later, the ion trap designs, met this need.

Capabilities of GC–MS

GC–MS combines the advantages of both techniques: the high resolving power and the speed of analysis of GC is retained, while the MS provides both positive identification and quantitative analyses down to the ppb level. Mass ranges of 10 to 600 Daltons are common for low cost systems, and up to 1000 Daltons for more expensive systems. Costs range from $40K up to $75K for simple bench-top systems.

Limitations of GC–MS Systems

GC–MS instruments are a capital expense item; they are more complicated to operate than a GC, and there is a lack of skilled GC–MS operators. Few colleges train students on GC–MS systems due to both a lack of systems for teaching purposes and the lack of expertise of many college professors.

Data Analysis

A typical capillary chromatogram of a hydrocarbon sample run on GC–MS has the same appearance as it would with an FID (see Fig. 10.9). Note the narrow peak widths, typically around 1 second or less at half height. This means that the MS system must scan the GC peak about 10 times per second in order to get a good mass spectrum.

Figure 10.10 shows a proposed mechanism for the fragmentation of n-hexane (peak 4 in Fig. 10.9) in the ion source of an GC–MS system. An electron strikes the parent molecule, ejecting one electron and generating the molecular ion (m/z = 86). This species is not stable, however, and rapidly decomposes to more stable fragments; in this case m/z of 71, 57, 43 and 29 Daltons. That fragment with the highest abundance, m/z = 57,

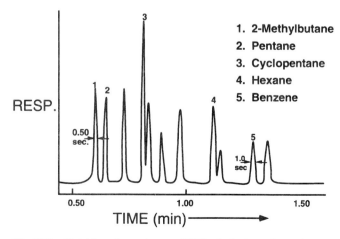

Fig. 10.9. Total ion chromatogram (TIC) of a hydrocarbon mixture.

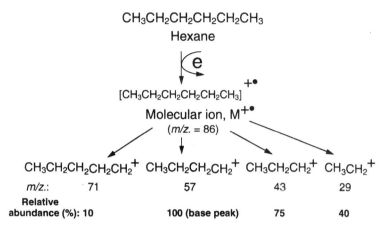

Fig. 10.10. Fragmentation of hexane in a mass spectrometer (EI).

is called the base peak, and the data system plots it as 100% of the spectrum scale. Other peaks are plotted relative to the base peak, and the result is a typical mass spectrum of n-hexane (see Fig. 10.11).

Data can be plotted in two ways; either as a total scan (TIC—Total Ion Chromatogram) or as a small number of individual ions (SIM—Selected Ion Monitoring) characteristic of a particular compound (see Fig. 10.12). A total ion chromatogram is used to identify unknown compounds. A specified mass range is scanned—for example 40 to 400 Daltons. All peaks are reported so the mass spectra can be retrieved from the computer and

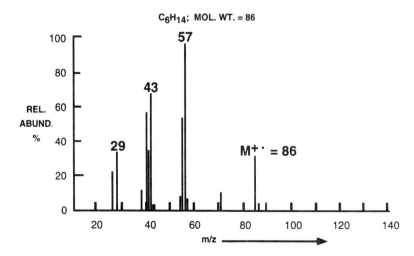

Fig. 10.11. Mass spectrum of hexane (EI).

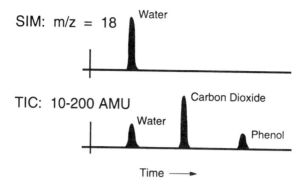

Fig. 10.12. Comparison of total ion chromatogram (TIC) and selected ion monitoring (SIM).

be used to identify each peak. The computer data base rapidly compares each unknown mass spectrum with over 150,000 reference spectra in its library files. Matching of spectra requires only a few seconds with the latest data systems, achieving the desired qualitative analysis. The data acquisition rate necessary to scan all ions in the selected range is slow; sensitivity is limited, and usually quantitation is not optimal (too few data points).

In selected ion monitoring, however, only a small number of ions (typically as many as 6) are monitored. There is a faster data acquisition rate during the lifetime of the GC peak (~ 1 second), so quantitative data is better and sensitivity is greatly improved. SIM cannot be used for qualitative analyses (not all masses are scanned), but it is the best mode for trace analysis of targeted compounds, often down to the ppb level.

A newer version of the ion trap [7] allows it to be operated in such a way that the original fragments from the ionization process are again exposed to energetic gas molecules causing a secondary fragmentation from the collision-induced dissociation (CID). The result is similar to that obtained by a dual stage mass spectrometer (usually referred to as MS/MS), and provides even higher selectivity. This mode of operation is called selected reaction monitoring (SRM) and is available in some newer ion trap GC–MS instruments.

CHIRAL ANALYSIS BY GC

Chiral separation by either GC or HPLC is an essential step in the synthesis, characterization and the utilization of chiral compounds (drugs, pesticides, flavors, pheromones, etc.). As an understanding of the significance of chirality on biological activity increases, legislation regulating chiral compounds becomes more widespread and stringent, and the need for high resolution separation techniques increases. Chiral separation by capillary GC provides high efficiency, sensitivity, and speed of analysis, but is limited by the need for volatility. Combining chiral phases into polysiloxanes has resulted in increased temperature stability.

GC separation of enantiomers can be performed either *direct* (use of a chiral stationary phase, CSP) or *indirect* (off-column conversion into diastereomeric derivatives and separation by non-chiral stationary phases). The direct method is preferred as being simpler and minimizing losses during sample preparation. The key, of course, is to find a chiral stationary phase with both selectivity and temperature stability.

There are three main types of chiral GC stationary phases: (1) chiral amino acid derivatives [8–10]; (2) chiral metal coordination compounds [11]; and (3) cyclodextrin derivatives [12–14]. The cyclodextrin phases have proven to be the most versatile for gas chromatography.

SPECIAL SAMPLING METHODS

Headspace Sampling

Samples containing nonvolatile materials present problems for gas chromatography. The non-volatiles cannot be injected into the GC as they will rapidly plug up the injection port and may destroy the GC column. Common sample preparation steps for isolating the volatiles from the nonvolatile sample matrix include liquid-liquid extraction, solid phase extraction (SPE), solid phase microextraction (SPME), supercritical fluid extraction (SFE), and headspace. SPE and SPME for GC are discussed in a recent review article by Penton [15].

Headspace sampling is probably the simplest and easiest technique. A brief introduction to the topic has been published by Hinshaw [16], and a complete coverage of the theory and practice has just appeared [17]. The sample (liquid or solid) is placed in a sealed vial and heated to a predetermined temperature for a fixed period of time. Volatile components of the sample partition between the gas and sample phases, usually reaching equilibrium. Residual monomers diffuse only slowly from some highly crosslinked polymers, so sufficient time must be allowed for the vaporization from these samples.

A portion of the volatiles in the gas phase (headspace) is removed and injected into the gas chromatograph. The simplest transfer technique is to use a heated gas-tight syringe and sample manually from convenient containers (paint cans for arson residues, beverage bottles for flavors and fragrances, etc.).

Better precision and accuracy are obtained from automated headspace samplers, where fixed volume vials are thermostated at controlled temperatures and times, and the headspace automatically transferred to the GC through inert (fused silica) heated transfer lines. Under controlled conditions, the headspace concentration of analytes is proportional to the original concentration in the sample. Classical quantitative calibration procedures such as internal standard, external standard and standard addition can be used to improve precision.

Solid Phase Microextraction (SPME)

SPME is a recent sample prep technique for trace analysis by GC [18]. It is a simple, solvent-free method which uses a nonpolar fiber (usually dimethylpolysiloxane) to extract analytes from a polar matrix (usually aqueous). A fused silica fiber is coated with a thin film (7, 30, or 100 μm) of stationary phase. The small size and cylindrical geometry allow easy incorporation of the coated fiber into an ordinary GC syringe. The coated fiber is exposed to the sample matrix or to the headspace, and analytes are adsorbed (extracted) from the sample matrix. After the fiber is removed from the sample, it is transferred to the heated inlet of a GC system and the analytes are thermally desorbed for analysis. The technique works well with trace amounts of nonpolar and semipolar analytes in water.

Figure 10.13 shows schematically the two main steps, (*a*) extraction (adsorption; Steps A–C) from the sample matrix and (*b*) desorption (Steps D–F) into the GC. *Step A:* the syringe with the fiber inside pierces the septum of a sample vial. Most often the sample is a liquid matrix, or a solution of a solid sample. *Step B:* the small fiber is extended out of the syringe and immersed in the solution for a predetermined time (stirring

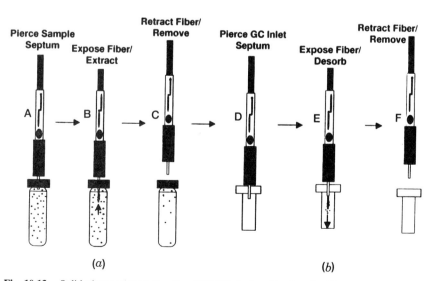

(*a*)   (*b*)

Fig. 10.13. Solid phase micro extraction (SPME) device. Reprinted with permission of Supelco, Bellefonte, PA 16823 USA from their 1997 catalog, p. 347.

helps) to reach an analyte equilibrium between the solid fiber and the liquid sample matrix. *Step C:* the fiber is retracted into the more mechanically stable syringe and removed from the sample vial. These steps can be done manually or automatically by modified GC autosamplers.

Step D: the syringe now pierces the septum of a GC and the fiber is exposed to the heated injection port where thermal desorption occurs (*Step E*). The fiber can be left in the injection port during the GC analysis in order to thoroughly clean the fiber for the next run. *Step F,* the fiber is retracted inside the syringe, the syringe is removed from the injection port, and process is ready for the next sample.

There are several advantages of this technique: 1) no organic solvents are used for extraction; 2) the technique is simple, and in the manual mode, costs are small; 3) the technique shows acceptable precision, 10–15% RSD at trace levels (down to 100 ppb).

There are both nonpolar and polar coatings available. Dimethylpolysiloxane is the most popular nonpolar one, and the 7 μm thin film is best for high molecular weight analytes; the 30 μm film is preferable for midrange (pesticides), and the 100 μm film for volatiles. Extraction efficiency depends on many factors: the chemical nature of the analyte, the sample matrix and the polymer coating; the extraction time and temperature; the degree of stirring and the analyte concentration. The desorption step depends primarily on the injection port temperatures, the volatility of the analyte, and the film thickness.

Volatile samples can be extracted by simply exposing the fiber to the headspace above a sample (liquid or solid matrix). Solid samples can be handled either by the headspace technique or by solution in an appropriate solvent. In some cases, the addition of salts ("salting out") increases the extraction efficiency of nonpolars from aqueous solutions.

Figure 10.14 shows a typical SPME application. Twenty common pesticides and an internal standard at the 200 ppt level in drinking water are extracted by a 100 μm polydimethylsiloxane fiber immersed for 15 min. in 4 mL of drinking water. A recent review [19] and a new book [20] can be consulted for other applications.

DERIVATIZATION

There are many reasons for performing chemical reactions on samples to form derivatives. Two reasons that are beneficial for gas chromatographic analysis are: the derivatization causes a nonvolatile sample to become

---

Fig. 10.14.   Chlorinated pesticide separation by SPME/WCOT GC. Reprinted with permission of Supelco, Bellefonte, PA 16823 USA from their 1997 catalog, p. A84.

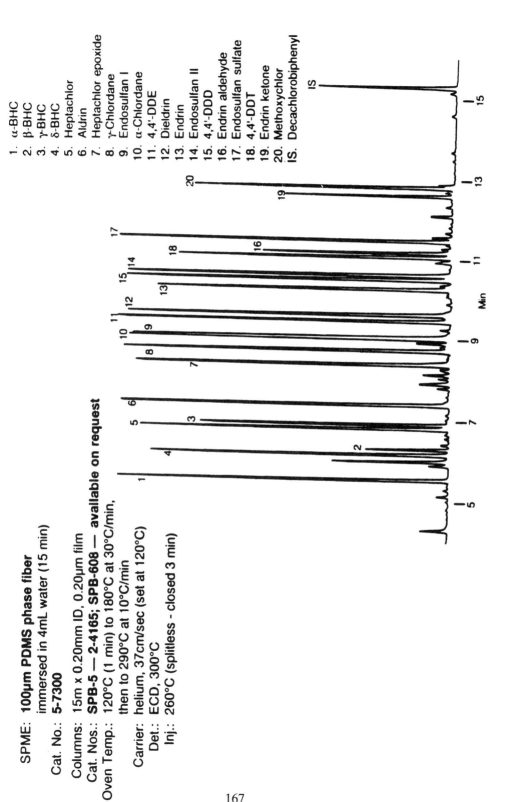

SPME: **100µm PDMS phase fiber**
immersed in 4mL water (15 min)
Cat. No.: **5-7300**

Columns: 15m x 0.20mm ID, 0.20µm film
Cat. Nos.: **SPB-5 — 2-4165; SPB-608 — available on request**
Oven Temp.: 120°C (1 min) to 180°C at 30°C/min,
then to 290°C at 10°C/min
Carrier: helium, 37cm/sec (set at 120°C)
Det.: ECD, 300°C
Inj.: 260°C (splitless - closed 3 min)

1. α-BHC
2. β-BHC
3. γ-BHC
4. δ-BHC
5. Heptachlor
6. Aldrin
7. Heptachlor epoxide
8. γ-Chlordane
9. Endosulfan I
10. α-Chlordane
11. 4,4'-DDE
12. Dieldrin
13. Endrin
14. Endosulfan II
15. 4,4'-DDD
16. Endrin aldehyde
17. Endosulfan sulfate
18. 4,4'-DDT
19. Endrin ketone
20. **Methoxychlor**
IS. **Decachlorobiphenyl**

167

TABLE 10.1 Guide to Derivatization[a]

Functional Group	Method	Derivatives
Acids	Silylation	$RCOOSi(CH_3)_3$
	Alkylation	RCOOR′
Alcohols and phenols– unhindered and moderately hindered	Silylation Acylation Alkylation	$R-O-Si(CH_3)_3$ $R-O-\overset{\overset{O}{\|\|}}{C}-PFA$ $R-O-R′$
Alcohols and phenols– highly hindered	Silylation Acylation Alkylation	$R-O-Si(CH_3)_3$ $R-O-\overset{\overset{O}{\|\|}}{C}-PFA$ R O R′
Amines (1° & 2°)	Silylation Acylation Alkylation	$R-N-Si(CH_3)_3$ $R-N-\overset{\overset{O}{\|\|}}{C}-PFA$ $R-N-R′$
Amines (3°)	Alkylation	PFB Carbamate
Amides	Silylation (a) Acylation (b) Alkylation (c)	(a) $R\overset{\overset{O}{\|\|}}{C}-NHSi(CH_3)_3$ (unstable) (b) $R\overset{\overset{O}{\|\|}}{C}-NH-\overset{\overset{O}{\|\|}}{C}$ PFA (c) $R\overset{\overset{O}{\|\|}}{C}-NHCH_3$
Amino Acids	Esterification/Acylation Silylation (a) Acylation + Silylation (b) Alkylation (c)	(a) $RCHCOOSi(CH_3)_3$ $\|$ $N-Si (CH_3)_3$ (b) $RCHOOSi(CH_3)_3$ $\|$ $N-TFA$ (c) $RCHCOOR′$ $\|$ $NHR′$
Catecholamines	Acylation + Silylation (a) Alcylation (b)	(a) benzene ring with R—N(H)—HFB, O Si(CH_3)_3, O Si(CH_3)_3 (b) benzene ring with R—N(H)—HFB, OHFB, OHFB

TABLE 10.1 *(Continued)*

Functional Group	Method	Derivatives
Carbohydrates and sugars	Silylation (a) Acylation (b) Alkylation (c)	(a) $\quad O\ Si(CH_3)_3$ $\qquad \mid$ $\qquad -(CH_2)_X-$ (b) $\quad$ OTFA $\qquad \mid$ $\qquad -(CH_2)_X-$ (c) $\quad$ OR $\qquad \mid$ $\qquad -(CH_2)_X-$
Carbonyls	Silylation	$TMS-O-N{=}C{<}$
	Alkylation	$CH_3-O-N{=}C{<}$

[a] Courtesy of Regis Chemical.

[b] Abbreviations: TMS = Trimethyl silyl; PFA = Perfluoroacyl; TFA = Trifluoroacetyl; HFB = Heptafluorobutyryl

volatile, or it improves the detectability of the derivative. This discussion mainly concerns the improvement of volatility which can prevent column fouling, a common problem for bio-separations [21]. In addition, derivatization often has a desirable secondary effect since the derivatives may also be more thermally stable.

Some monographs on derivatization are listed in the References [22–25] along with some relevant publications by laboratory supply houses [26–28].

Classification of Reactions

The reactions to produce volatile derivatives can be classified as silylation, acylation, alkylation, and coordination complexation. Examples of the first three types are included in Table 10.1 which is organized by functional groups including: carboxylic acid, hydroxyl, amine, and carbonyl. Amines require special consideration even if they are volatile. Their strong tendency to hydrogen-bond often makes it difficult to elute them from a GC column. Consequently, amines often have to be derivatized whether they are volatile or not. A review of this subject has appeared recently [29].

The fourth reaction type, coordination complexation, is used with metals, and typical reagents are trifluoroacetylacetone and hexafluoroacetylacetone [30]. Drozd [31] has reviewed this field and provided over 600 references and no further discussion is presented here.

Silylation reactions are very popular and need further description. A variety of reagents are commercially available, and most are designed to introduce the trimethylsilyl group into the analyte to make it volatile. A typical reaction is the one between bis-trimethylsilylacetamide (BSA) and an alcohol:

$$R\text{—}OH + CH_3\text{—}C\overset{O\text{—}Si(CH_3)_3}{\underset{N\text{—}Si(CH_3)_3}{}} \longrightarrow R\text{—}O\text{—}Si(CH_3)_3 + CH_3\text{—}C\overset{O\text{—}H}{\underset{N\text{—}Si(CH_3)_3}{}} \qquad (4)$$

A closely related reagent contains the trifluoroacetamide group and produces a more volatile reaction by-product (not a more volatile derivative); the reagent is bis(trimethylsilyl)-trifluoroacetamide (BSTFA). The order of reactivity of the silylation reagents[a] is:

TSIM ≥ BSTFA ≥ BSA ≥ MSTFA ≥ TMSDMA ≥
TMSDEA ≥ TMCS ≥ HMDS

In general, the ease of reaction follows the order [10]:

alcohols ≥ phenols ≥ carboxylic acid ≥ amines ≥ amides

If a solvent is used, it is usually a polar one; the bases DMF and pyridine are commonly used to absorb the acidic by-products. An acid catalyst such as trimethylchlorosilane (TMCS) and heating are sometimes needed to speed up the reaction.

Methods of Derivatization

The methods of derivatization can be divided into several categories: pre- and post-column methods and off-line and on-line methods. For example, the formation of volatile derivatives for GC is usually prepared off-line in separate vials before injection into the chromatograph (precolumn). There are a few exceptions where the reagents are mixed and injected together; the derivatization reaction occurs in the hot GC injection port (on-line).

Precolumn reactions that do not go to completion will produce mixtures that are more complex than the starting sample. As a result, excess reagent is usually used to drive the reaction to completion, thus leaving an excess of the reagent in the sample. Unless a prior separation step is used, the chromatographic method must be designed to separate these additional impurities. When performed off-line, the precolumn techniques can be used with slow reactions and heated to provide better quantitative results.

Improved detectivity usually arises from the incorporation of a *chromophore* into the analytes. In GC, one example is the incorporation into the

---

[a] The reagent names not identified in the text are:

TSIM = Trimethylsilylimidazole
MSTFA = N-Methyl-trimethylsilyltrifluoroacetamide
TMSDMA = Trimethylsilyldimethylamine
TMSDEA = Trimethylsilyldiethylamine
HMDS = Hexamethyldisilazane

analytes of functional groups that will enhance their detectivity by a selective detector such as the ECD. The purpose of forming the derivatives is to improve the limit of detection or the selectivity or both. Another example is the use of deuterated reagents to form derivatives that can be easily distinguished by their higher molecular weight when analyzed by GC–MS.

Summary

Derivatization offers one method for analyzing relatively nonvolatile samples by GC, but there are those who feel that it would be better to perform such analyses by other means, so one has to decide for him/herself. At a minimum, the formation of derivatives inserts an extra step or steps into an analytical procedure, raising the possibility of additional errors and requiring extra method validation.

Incorporation of derivatization into a quantitative method of analysis may be facilitated by the use of an internal standard (see Chapter 8). In that case, the internal standard should be added to the sample before the derivatization is performed.

REFERENCES

(GC–MS)

1. Watson, J. T., *Introduction to Mass Spectrometry,* 2nd ed., Lippincott-Raven, New York, 1985.
2. McLafferty, F. W., *Interpretation of Mass Spectra,* third edition, University Science Books, Mill Valley, CA, 1980.
3. March, R. E., and Hughes, R. J., *Quadrupole Storage Mass Spectrometry,* Wiley-Interscience Publication, New York, 1989.
4. Message, G. M., *Practical Aspects of Gas Chromatography/Mass Spectrometry,* Wiley, New York, 1984.
5. Kitson, F. G., Larsen, B. S., and McEwen, C. N., *Gas Chromatography and Mass Spectrometry: A Practical Guide,* Academic Press, San Diego, 1996.
6. Gohlke, R. S., *Anal. Chem.* **31,** 534 (1959).
7. Sheehan, T. L., *Am. Lab.,* **28,** (17), 28V (1996).

(Chiral Separations)

8. Gil-Av, E., *J. Mol. Evol.,* **6,** 131 (1975).
9. Bayer, E., and Frank, H., *ACS Symposium Series #121,* American Chemical Society, Washington, D.C., 1980, p. 34.
10. Konig, W. A., *Enantioselective Gas Chromatography with Modified Cyclodextrins,* Huthig, Heidelberg, 1992.
11. Konig, W. A., *J. High Resolut. Chromatogr.,* **5,** 588 (1982).
12. Schurig, F. V., and Nowotny, H. P., *Angew. Chem. Int. Ed. Engl.,* **29,** 939 (1990).
13. Schurig, V., *J. Chromatogr.,* **441,** 135 (1988).
14. Konig, W. A., *Kontakte* **2,** 3 (1990).

(Special Sampling Methods)

15. Penton, Z. E. *Advances in Chromatography* (37) P. R. Brown and E. Grushka, Eds., Dekker, NY, 1997, Chpt. 5.

16. Hinshaw, J. V., *LC-GC,* **8,** 362 (1990).

17. Bruno, K., and Ettre, L. S., *Static Headspace Gas Chromatography: Theory and Practice,* Wiley, NY, 1997.

18. Zhang, Z., Yang, M. J., and Pawliszyn, J., *Anal. Chem.,* **66,** 844A (1994).

19. Eiseret, R., and Levsen, K., *J. Chromatogr. A,* **733,** 143 (1996).

20. Pawliszyn, J., *Solid Phase Microextraction: Theory and Practice,* Wiley, NY, 1997.

21. Rood, D., *LC-GC,* **8,** 216 (1990).

(Derivatization)

22. Pierce, A. E., *Silylation of Organic Compounds,* Pierce Chemical, Rockford, IL, 1968.

23. Knapp, D. R., *Handbook of Analytical Derivatization Reactions,* Wiley, New York, 1979.

24. Drozd, J., *Chemical Derivatization in Gas Chromatography,* Elsevier, Amsterdam, 1981.

25. Blau, K., and Halket, J. M., *Handbook of Derivatives for Chromatography,* Wiley, New York, 1993.

26. *Derivatization for Gas Chromatography, GC Derivatization Reagents,* and *Derivatization Wall Chart,* Regis Technologies, Inc., Morton Grove, IL, 60053.

27. *Handbook of Derivatization,* Pierce Chemical Co., Rockford, IL, 61105.

28. Simchen, G., and Heberle, J., *Silylating Agents,* Fluka Chemical Corp., Ronkonkoma, NY, 1995.

29. Kataoka, H., *J. Chromatogr. A,* **733,** 19 (1996).

30. Mosier, R. W., and Sievers, R. E., *Gas Chromatography of Metal Chelates,* Pergamon Press, Oxford, 1965.

31. Drozd, J., *J. Chromatogr.,* **113,** 303 (1975).

11. Troubleshooting GC Systems

The following pages have been inserted to help the chromatographer interpret the different peak shapes encountered in gas chromatography. The various chromatograms obtained are the result of our own experiences combined with a thorough literature search.

The injection point on each chromatogram is shown by a tick mark on the baseline as shown in example 1. The time axis runs from left to right (see arrow).

SYMPTOM	POSSIBLE CAUSE	CHECKS AND/OR REMEDY
1. No peaks.	1a. Main power off, fuse burned out.	1a. Plug in system, check fuses.

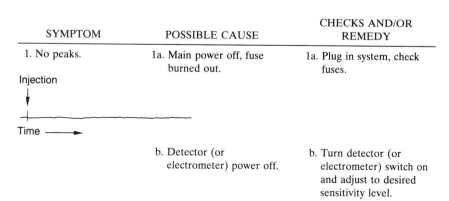

	b. Detector (or electrometer) power off.	b. Turn detector (or electrometer) switch on and adjust to desired sensitivity level.

SYMPTOM	POSSIBLE CAUSE	CHECKS AND/OR REMEDY
	c. No carrier gas flow.	c. Turn carrier gas flow ON and adjust to proper setting. If carrier lines are obstructed, remove obstruction. Replace carrier-gas tank if empty.
	d. Integrater/data system improperly connected; not turned on; not grounded.	d. Connect systems as described in manual. Remove any jumper lines connecting either input connection to ground or shield.
	e. Injector temperature too cold. Sample not being vaporized.	e. Increase injector temperature. Check with volatile sample such as air or acetone.
	f. Hypodermic syringe leaking or plugged up.	f. Squirt acetone from syringe onto paper; if no liquid comes out, then replace the syringe.
	g. Injector septum leaking.	g. Replace injector septum.
	h. Column connections loose.	h. Use leak detector, check leaks, tighten column connections.
	i. Flame out (FID only).	i. Inspect flame; check to see if water vapor condenses on mirror, light if necessary.
	j. No cell voltage being applied to detector (all ionization detectors).	j. Place CELL VOLTAGE in ON position. Also check for bad detector cables. Measure voltage with a voltmeter per instruction manual.
	k. Column temperature too cold. Sample condensing on column.	k. Inject volatile compound like air or acetone, increase column temperature.
2. Poor sensitivity with normal retention time.	2a. Attenuation too high.	2a. Reduce attenuation.
	b. Insufficient sample size.	b. Increase sample size; check syringe.
	c. Poor sample injection technique.	c. Review sample injection techniques.

SYMPTOM	POSSIBLE CAUSE	CHECKS AND/OR REMEDY
	d. Syringe or septum leaking when injecting.	d. Replace syringe or septum.
	e. Carrier gas leaking.	e. Find and correct leak; usually retention time will also change.
	f. Thermal conductivity response low.	f. Use higher filament current; He or H_2 carrier gas.
	g. FID response low.	g. Optimize both air and H_2 flow rate; use N_2 to make up gas.
3. Poor sensitivity with increased retention time.	3a. Carrier gas flow rate too low.	3a. Increase carrier gas flow. If carrier gas lines are obstructed, locate and remove obstruction.
	b. Flow leaks downstream of injector; usually at column inlet.	b. Locate flow leak and correct.
	c. Injector septum leaking continuously.	c. Replace injector system.
4. Negative peaks.	4a. Integrator/data system improperly connected. Input leads reversed.	4a. Connect system as described in manual.
	b. Sample injected in wrong column.	b. Inject sample in proper column; only on dual column systems!
	c. MODE switch in wrong position (ionization detectors).	c. Insure MODE switch is in correct position for column being used as analytical column.
	d. POLARITY switch in wrong position (thermal conductivity detector).	d. Change POLARITY switch.
5. Irregular baseline drift when operating isothermally.	5a. Poor instrument location.	5a. Move instrument to a different location. Instrument should not be placed directly under heater or air conditioner blower, or any other place where it is subject to excessive drafts and ambient temperature changes.
	b. Instrument not properly grounded.	b. Insure instrument and data system connected to good earth ground.

SYMPTOM	POSSIBLE CAUSE	CHECKS AND/OR REMEDY
	c. Column packing bleeding.	c. Stabilize column as outlined in instrument manual. Some columns are impossible to stabilize well at the desired operating conditions. These columns will always produce some baseline drift, particularly when operating at high sensitivity conditions.
	d. Carrier gas leaking.	d. Locate leak and correct.
	e. Detector block contaminated.	e. Clean detector block. Raise temperature and bake out detector over night.
	f. Detector base contaminated (ionization detectors).	f. Clean detector base. See instrument manual.
	g. Poor carrier gas regulation.	g. Check carrier gas regulator and flow controllers to insure proper operation. Make sure carrier gas tank has sufficient pressure.
	h. Poor H_2 or air regulation (FID only).	h. Check H_2 and air flow to insure proper flow rate and regulation.
	i. Detector filaments defective (TC detector only).	i. Replace TC detector assembly or filament.
	j. Electrometer defective (ionization detectors).	j. See instrument manual on electrometer troubleshooting.
6. Sinusoidal baseline drift.	6a. Detector oven temperature controller defective.	6a. Replace detector oven temperature controller, and/or temperature sensing probe.
	b. Column oven temperature defective.	b. Replace oven temperature control module, and/or temperature sensing probe.
	c. OVEN TEMP °C control on main control panel set too low.	c. Set OVEN TEMP °C control to higher setting. Must be set higher than highest desired operating.

SYMPTOM	POSSIBLE CAUSE	CHECKS AND/OR REMEDY
		temperature of the column oven.
	d. Carrier gas flow regulator defective.	d. Replace carrier gas flow regulator; sometimes higher pessure provides better control.
	e. Carrier gas tank pressure too low to allow regulator to control properly.	e. Replace carrier gas tank.
7. Constant baseline drift in one direction when operating isothermally.	7a. Detector temperature increasing (decreasing).	7a. Allow sufficient time for detector to stabilize after changing its temperature. Particularly important with TC detector. Detector block will lag the indicated temperature somewhat because of its large mass.
	b. Flow leak down stream of column effluent end (TC detector only).	b. A very small diffusion leak will allow a small amount of air to enter the detector at a constant rate. This in turn will oxidize the effected elements at a constant rate while slowly changing their resistance. Locate the leak and correct. These are very often very slight leaks, and difficult to find. Use high carrier gas pressure (60–70 psig) is necessary.
	c. Defective detector filaments (TC detector).	c. Replace detector or filaments.
8. Rising baseline when temperature programming.	8a. Increase in column "bleed" when temperature rises.	8a. Use less liquid phase and lower temperature. If possible, use more temperature stable liquid phase.
	b. Column(s) contaminated.	b. 1) Bake out column overnight. 2) Break off first 10 cm of column inlet.

SYMPTOM	POSSIBLE CAUSE	CHECKS AND/OR REMEDY
9. Irregular baseline shifting when temperature programming.	9a. Excessive column "bleeding" from well conditioned columns.	9a. Use less liquid phase and low temperatures. Use different columns.
	b. Columns not properly conditioned.	b. Condition columns as outlined in instruction manual.
	c. Column(s) contaminated.	c. See 8b.
10. Baseline cannot be zeroed.	10a. Zero on data system improperly set.	10a. Reset zero. Short system input with piece of wire and adjust to zero. See system instruction manual.
	b. Detector filaments out of balance (TC detector).	b. Replace detector.
	c. Excessive signal from column "bleed" (especially FID).	c. Use different column with less "bleed." Use lower column temperature.
	d. Dirty detector (FID and EC).	d. Clean detector base and head assemblies.
	e. Data system improperly connected.	e. Connect system as described in instrument manual. Remove any jumper lines connecting either system input connection to ground or shield.
11. Sharp "spiking" at irregular intervals.	11a. Quick atmospheric pressure changes from opening and closing doors, blowers, etc.	11a. Locate instrument to minimize problem. Also do not locate under heater or air conditioner blowers.
	b. Dust particles or other foreign material burned in flame (FID only).	b. Take care to keep detector chamber free of glass wool, maranite, molecular sieve (from air filter), dust particles, etc. Blow out or vacuum detector to remove dust.
	c. Dirty insulators and/or connectors (Ionization detectors).	c. Clean insulators and connectors with residue free solvent. Do not touch with bare fingers after cleaning.
	d. High line voltage fluctuations.	d. Use separate electrical outlet; use stabilized transformer.

SYMPTOM	POSSIBLE CAUSE	CHECKS AND/OR REMEDY
12. High background signal (noise).	12a. Contaminated column or excessive "bleed" from column.	12a. Recondition column (see 8b.)
	b. Contaminated carrier gas.	b. Replace or regenerate carrier gas filter. Regenerate filter by heating to about 175–200°C and purging overnight with dry nitrogen.
	c. Carrier gas flow rate too high.	c. Reduce carrier gas flow rate.
	d. Carrier gas flow leak.	d. Locate leak and correct.
	e. Loose connections.	e. Make sure all interconnecting plug and screw connections are tight. Make sure modules are properly seated in their plug-in connectors.
	f. Bad ground connection.	f. Insure all ground connections are tight and connected to a good earth ground.
	g. Dirty switches.	g. Locate dirty switch, spray with a contact cleaner and rotate switch through its positions several times.
	h. Dirty injector.	h. Clean injector tube and replace septum.
	i. Dirty crossover block from column oven to detector oven.	i. Clean crossover block.
	j. Dirty detector (TC detector).	j. Clean detector block.
	k. Defective detector filaments (TC detector).	k. Replace detector assembly.
	l. Hydrogen flow rate too high or too low (FID detector).	l. Adjust hydrogen flow rate to proper level.
	m. Air flow too high or too low (FID detector).	m. Adjust air flow rate to proper level.

APPENDIX I. LIST OF SYMBOLS AND ACRONYMS

A	Peak area
A_s	Surface area of stationary phase in column
d	Distance between maxima of two adjacent peaks
d_c	Column inside diameter
d_f	Thickness of liquid phase
d_p	Particle diameter
D	Minimum detectability of a detector
D	Diffusion coefficient in general
D_G	Diffusion coefficient in the gas phase
D_L	Diffusion coefficient in liquid stationary phase
D_M	Diffusion coefficient in the mobile phase
D_S	Diffusion coefficient in the stationary phase
ECD	Electron capture detector
f	Relative detector response factor
F	Mobile-phase flow rate, measured at column outlet under ambient conditions with a wet flow meter
F_c	Mobile-phase flow rate corrected
FID	Flame ionization detector
FTIR	Fourier Transform infrared
GC	Gas chromatography
GLC	Gas-liquid chromatography
GLPC	Gas-liquid partition chromatography
GSC	Gas-solid chromatography
H	Plate height (HETP)
$\mathcal{H}$	Enthalpy
HETP	Height equivalent to one theoretical plate
I	Retention index; Kovats
j	Mobile phase compression (compressibility) correction factor
k	Retention factor (capacity factor)
K_c	Distribution constant in which the concentration in the stationary phase is expressed as mass of substance per volume of the phase
L	Column length
LC	Liquid chromatography
MDQ	Minimum detectable quantity
MS	Mass spectroscopy
N	Noise of a detector
N	Plate number (number of theoretical plates)
OT	Open tubular (column)
p	Pressure in general; partial pressure
P_i	Inlet pressure
P_o	Outlet pressure
p^0	Equilibrium vapor pressure
PLOT	Porous-layer open tubular (column)
r_c	Inside column radius

R	Retardation factor in column chromatography; fraction of a sample component in a mobile phase
$\mathcal{R}$	Gas constant
R_s	Peak resolution
S	Detector sensitivity
SCOT	Support-coated open tubular (column)
t	Time in general
t_M	Mobile-phase hold-up time; it is also equal to the retention time of an unretained compound
t_R	Peak elution time
t_R'	Adjusted retention time
t_R^o	Corrected retention time
T	Temperature in general (always in Kelvin)
T'	Significant temperature (in PTGC)
T_c	Column temperature
TCD	Thermal conductivity detector
TF	Tailing factor
u	Mobile-phase velocity
$\bar{u}$	Average linear carrier gas velocity
V	Volume in general
V_g	Specific retention volume at 0°C
V_G	Interparticle volume of column in GC
V_L	Liquid-phase volume
V_M	Mobile-phase hold-up volume; also equal to the retention volume of an unretained compound
V_M^0	Corrected gas hold-up volume
V_M	Volume of mobile phase in column
V_N	Net retention volume
V_R	Total retention volume
V_R'	Adjusted retention volume
V_R^o	Corrected retention volume
V_S	Volume of stationary phase in column
w_b	Peak width at base
w_h	Peak width at half height
WCOT	Wall coated open tubular (column)
z	Number of carbon atoms of an *n*-alkane eluted before the peak of interest
$(z + 1)$	Number of carbon atoms of an *n*-alkane eluted after the peak of interest

Greek symbols

α	Separation factor (relative retardation)
β	Phase ratio
γ	Activity coefficient
λ	Packing factor (Rate Equation)

μ	Velocity of solute
σ	Standard deviation of a Gaussian peak
σ^2	Variance of a Gaussian peak
τ	Time constant (Detector)
ω	Packing factor (Rate Equation)

APPENDIX II.   GUIDELINES FOR SELECTING
CAPILLARY COLUMNS

I. Length
  A. Rule: Use shortest useful column
    1. Save time
    2. Cheaper
    3. Reduced side effects (reduced residence time)
    4. If more R_S required, consider reducing d_f and/or i.d.
II. Internal Diameter
  A. Megabore (0.53mm i.d.) preferred when high carrier flow rate desired
    1. Simple direct injection techniques
    2. Primitive equipment including dead volumes, cold spots, active materials, parts that cannot be cleaned
    3. Sample transfer from absorbent filters (head space, SFC, SPE techniques)
  B. Medium size columns (0.25–0.35mm i.d.)
    1. Commonly used as good compromise
  C. Narrow columns (0.10mm i.d.) for increased separation efficiency and speed
    1. Shorter lengths are possible and faster analysis
    2. Limitations
      a. High split ratios necessary (500/1)
      b. Limited trace analyses
      c. High carrier gas pressures required
      d. Equipment and manipulation more critical
III. Film Thickness
  A. Advantages of thick films
    1. Increased retention; frequently essential for volatiles; film thickness may replace column length
    2. Increased capacity; important for GC/MS or FTIR
    3. Elution shifted to higher temperature (all sample components see warmer column), resulting in reduced adsorption effects
  B. Advantages of thin films
    1. Maximum separation efficiency
    2. Elution shifted to lower temperature (sample sees cooler column)
    3. Faster analyses
IV. Stationary Phase
  A. Start with nonpolar phases like DB-1 or DB-5. More efficient, more inert and generally useful for most sample types. The non-polar character shows low solubility for polar compounds, thus allowing lower column temperatures to be used. This means better stability for thermolabile compounds.
  B. If greater selectivity is needed, try a more polar phase, OV-1701 or some version of Carbowax®.

V. Carrier Gases—Use H_2 or He (much faster than N_2)
  A. Advantages of H_2 over He
    1. Separation efficiency slightly higher
    2. Analysis time roughly 50% faster (isothermal only)
    3. Better sensitivity (sharper peaks)
    4. Columns regularly run at lower temperature, resulting in increased resolution and longer column life
  B. Limitations
    1. Potential hazard; may cause explosion if more than 5% in air and spark. Not recommended, especially not for GC/MS.

APPENDIX III.   GC: HOW TO AVOID PROBLEMS

I. Carrier Gas
   A. Use high purity gases, 99.9% minimum; 99.999 for GC–MS.
   B. Use a molecular sieve scrubber on *all gas cylinders* to remove H_2O and methane.
   C. Use of an O_2 scrubber on carrier gas line is essential for electron capture detector; recommended for high temperature capillary columns.
   D. Use He (or H_2) for TCD. N_2 is not sensitive (also it gives both $+$ and $-$ peaks).
   Use He or N_2 for FID.
   Use bone dry, O_2-free N_2 for ECD.
   E. Know the van Deemter (or Golay) plot for your column. $\bar{u}$ opt. is 12, 20, and 40 cm/sec for N_2, He and H_2 respectively. H vs. $\bar{u}$. Measure $\bar{u}$ daily (inject methane). $\bar{u} = L$ (cm)/t_M (secs).
II. Injectors
   A. Packed Column—use on-column injectors; more inert, lower temperature than off-column heated inlet. Use only a small piece of silanized glass wool. Don't pack the first few inches (see your manual) of the column to allow space for needle. Use the lowest possible inlet temperature which produces the least band broadening.
   B. Capillary Column
      1. Split—split in the range of 20/1 to 200/1. A good starting point is 50/1. Low split ratios give better sensitivity, but eventually lead to low resolution. For gas sample valves, purge and trap, and SFE interfaces increase split ratio until R_s is maximized. Use a fast injection technique, preferably with an autosampler.
      2. Splitless—
         a. Dilute sample in volatile solvent like hexane, iso-octane, or methylene chloride.
         b. Set column temperature at b.p. of solvent.
         c. Inject slowly, 1–5 μl, "hot needle" technique.
         d. Start temperature program; open split valve after 1 minute.
III. Columns
   A. Buy good columns from reliable manufacturers. Don't try to save a few dollars. Check out all columns regularly. Run your test mix; measure N, α, k, and R_s.
   B. Clean columns regularly. Best ways to clean a column:
      1. Bake out overnight;
      2. Cut off first 10 cm at least once a month.
      3. If necessary, take out column, rinse with solvents (only bonded phases), dry well, reinstall and condition slowly.
   Remember: Bad performance of a sample doesn't necessarily mean the column is bad; run a standard check on the column.

C. Capillary Columns
   1. Length—start with 25m; shorter columns are faster, longer columns have more plates (but are slow). It is better to use *thinfilm, small i.d.,* and *small sample sizes* to increase column efficiency.
   2. i.d.—start with 250 or 320 μm. Megabore (530 μm) are not as efficient; 100 μm require *very* small, *very* fast injections.
   3. Carrier gas—use He or H_2; N_2 is too slow.
   4. d_f—start with 0.2 or 0.5 μm. Thicker films for volatiles, but usually less efficient.

IV. Detectors
   A. Always use proper carrier gas; one of high purity.
   B. Use scrubbers to remove H_2O and light hydrocarbons.
   C. If necessary, use make-up gas. Essential for ECD and TCD; often increases sensitivity with FID.
   D. Keep the detector hot; avoid condensation of sample.

APPENDIX IV. CALCULATION OF SPLIT RATIO FOR SPLIT INJECTION ON OT COLUMNS

First, measure the flow rate out of the split vent using a suitable flow meter (soap film flow meter or electronic flow meter). Then, inject a 5 microliter sample of methane and record its retention time.

Calculate the average linear velocity of the carrier gas,

$$\bar{u} = L/t_M$$

where L = column length in cm. and t_M is the retention time for methane in seconds. The units for average linear velocity are usually cm/s.

To convert the velocity to a flow rate, F_c, the velocity must be multiplied by the cross sectional area of the column:

$$F_c = \bar{u} \, (\pi \, r^2) \, 60$$

Multiplying by 60 will convert the units to mL/min.

Finally, calculate the split ratio:

$$\text{Ratio} = \text{Split flow rate/column flow rate}$$

If both values are in mL/min, the units will cancel.

APPENDIX V. OPERATING CONDITIONS FOR CAPILLARY COLUMNS

Column Diameter (mm)	Length (meters)	Head Pressure (psi)	Linear Velocity (cm/sec)	Column Flow (mL/min)
0.10	10	38	30	0.13
0.25	12	4	30	0.75
	25	13	30	0.75
	50	30	30	0.75
0.32	25	11	30	1.35
	50	5	23	1.00
	50	25	30	1.35
0.53	10	2	50	7.00
	10	8	152	20.0
	50	24	227	30.0

APPENDIX VI. OV LIQUID PHASES PHYSICAL PROPERTY DATA. Reprinted with permission of Ohio Valley Specialty Company.

Name	Type	Structure	Solvent	Temp. Limit	Viscosity
OV-1	Dimethylsilicone Gum	$\left[\begin{smallmatrix}CH_3\\-Si-O-\\CH_3\end{smallmatrix}\right]_n$	Toluene	325–375°C	Gum
OV-101	Dimethylsilicone	$\left[\begin{smallmatrix}CH_3\\-Si-O-\\CH_3\end{smallmatrix}\right]_n$	Toluene	325–375	1,500
OV-3	Phenylmethyldimethylsilicone 10% Phenyl	$\left[\begin{smallmatrix}CH_3\\-Si-O-\\\phi\end{smallmatrix}\right]_n\left[\begin{smallmatrix}CH_3\\Si-O-\\CH_3\end{smallmatrix}\right]_m$	Acetone	325–375	500
OV-7	Phenylmethylsilicone 20% Phenyl	$\left[\begin{smallmatrix}CH_3\\-Si-O-\\\phi\end{smallmatrix}\right]_n\left[\begin{smallmatrix}CH_3\\Si-O-\\CH_3\end{smallmatrix}\right]_m$	Acetone	350–375	1,300
OV-11	Phenylmethylsilicone 35% Phenyl	$\left[\begin{smallmatrix}CH_3\\-Si-O-\\\phi\end{smallmatrix}\right]_n\left[\begin{smallmatrix}CH_3\\Si-O-\\CH_3\end{smallmatrix}\right]_m$	Acetone	325–375	500
OV-17	Phenylmethyldimethylsilicone 50% Phenyl	$\left[\begin{smallmatrix}CH_3\\-Si-O-\\\phi\end{smallmatrix}\right]_n$	Acetone	325–375	1,300
OV-61	Diphenyldimethylsilicone	$\left[\begin{smallmatrix}\phi\\-Si-O-\\\phi\end{smallmatrix}\right]_n\left[\begin{smallmatrix}CH_3\\Si-O-\\CH_3\end{smallmatrix}\right]_m$	Acetone	325–375	>50,000
OV-73	Diphenyldimethylsilicone Gum	$\left[\begin{smallmatrix}\phi\\-Si-O-\\\phi\end{smallmatrix}\right]_n\left[\begin{smallmatrix}CH_3\\Si-O-\\CH_3\end{smallmatrix}\right]_m$	Toluene	325–350	Gum
OV-22	Phenylmethyldiphenylsilicone	$\left[\begin{smallmatrix}\phi\\-Si-O-\\\phi\end{smallmatrix}\right]_n\left[\begin{smallmatrix}CH_3\\Si-O-\\\phi\end{smallmatrix}\right]_m$	Acetone	350–375	>50,000
OV-25	Phenylmethyldiphenylsilicone	$\left[\begin{smallmatrix}\phi\\-Si-O-\\\phi\end{smallmatrix}\right]_n\left[\begin{smallmatrix}CH_3\\-Si-O-\\\phi\end{smallmatrix}\right]_m$	Acetone	350–375	>100,000
OV-105	Cyanopropylmethyl-Dimethylsilicone		Acetone	275–300	1,500
OV-202	Trifluoropropylmethylsilicone	$\left[\begin{smallmatrix}CH_3\\-Si-O-\\C_2H_4\\CF_3\end{smallmatrix}\right]_n$	Chloroform	250–275	500
OV-210	Trifluoropropylmethylsilicone	$\left[\begin{smallmatrix}CH_3\\-Si-O-\\C_2H_4\\CF_3\end{smallmatrix}\right]_n$	Chloroform	275–350	10,000
OV-215	Trifluoropropylmethylsilicone Gum		Ethyl Acetate	250–275	Gum
OV-225	Cyanopropylmethyl- Phenylmethylsilicone	$\left[\begin{smallmatrix}CH_3\;\;CH_3\\-Si-O-Si-O-\\C_3H_6\;\;\phi\\C\equiv N\end{smallmatrix}\right]_n$	Acetone	250–300	9,000
OV-275	Dicyanoallylsilicone		Acetone	250–275	20,000
OV-330	Silicone Carbowax Copolymer		Acetone	250–275	500
OV-351	Polyglycol-Nitroterephthalic		Chloroform	250–275	Solid
OV-1701	Dimethylphenylcyano Substituted Polymer		Acetone	300–325	Gum

APPENDIX VI (*Continued*).

Wt. Avg, Average Mol Weight	△1 McReynolds Constants ①	②	③	④	⑤	⑥	⑦
>10⁶	16	55	44	65	42	32	23
3×10^4	17	57	45	67	43	33	23
2×10^4	44	86	81	124	88	55	46
1×10^4	69	113	111	171	128	77	66
7×10^3	102	142	145	219	178	103	92
4×10^3	119	158	162	243	202	112	105
4×10^4	101	143	142	213	174	99	86
8×10^5	40	86	76	114	85	57	39
8×10^3	160	188	191	283	253	133	132
1×10^4	178	204	208	305	280	144	147
	36	108	93	139	86	74	29
1×10^4	146	238	358	468	310	206	56
2×10^5	146	238	358	468	310	206	56
	149	240	363	478	315	208	56
8×10^3	228	369	338	492	386	282	150
5×10^3	629	872	763	1106	849	686	318
5×10^3	222	391	273	417	368	284	158
	335	552	382	583	540	–	–
	67	170	153	228	171	–	–

① Benzene ② Butanol ③ 2-Pentanone ④ Nitropropane ⑤ Pyridine ⑥ 2-Methyl-2-Pentanol ⑦ 2-Octyne

APPENDIX VII.   SOME PRESSURE CORRECTION FACTORS (j)

p_i/p_o	j	p_i/p_o	j
1.1	0.95	1.8	0.70
1.2	0.91	1.9	0.67
1.3	0.86	2.0	0.64
1.4	0.83	2.2	0.60
1.5	0.79	2.5	0.54
1.6	0.76	3.0	0.46
1.7	0.72	4.0	0.36

APPENDIX VIII. LIST OF SOME CHROMATOGRAPHIC SUPPLY HOUSES

1. Alltech Associates, Inc.
   2051 Waukegan Road
   Deerfield, IL 60015
   847-948-8600
2. Chrompack, Inc.
   1130 Route 202
   Raritan, NJ 08869
   800-526-3687
3. J&W Scientific
   91 Blue Ravine Road
   Folsom, CA 95630-4714
   916-985-7888
4. Restek Corporation
   110 Benner Circle
   Bellefonte, PA 16823-8812
   800-356-1688
5. Supelco Inc.
   Supelco Park
   Bellefonte, PA 16823
   800-359-3041

Plus instrument manufacturers such as:

1. Hewlett-Packard Company
   Analytical Products Group
   P.O. Box 9000
   San Fernando, CA 91341-9981
2. Perkin-Elmer
   761 Main Avenue
   Norwalk, CT 06859
3. Varian Analytical Instruments
   811 Hanson Way, B111
   Palo Alto, CA 94303-1031

APPENDIX IX. OTHER RESOURCES

Some Recent Books (see also the chapter references)

1. Baugh, P. J., *Gas Chromatography: A Practical Approach,* Oxford Univ. Press, 1994.

2. Braithwaite, A., and Smith, F. J., *Chromatographic Methods,* fifth edition, Chapman and Hall, 1996.

3. Fowlis, I. A., *Gas Chromatography,* second edition, ACOL Series, Wiley, Chichester, 1995.

4. Grant, D. W., *Capillary Gas Chromatography,* Wiley, New York, 1996.

5. Grob, R. L., Ed., *Modern Practice of Gas Chromatography,* third edition, Wiley, New York, 1995.

6. Hinshaw, J. V., and Ettre, L. S., *Introduction to Open-Tubular Column Gas Chromatography,* Advanstar, Cleveland, 1994.

7. Rood, D., *A Practical Guide to the Care, Maintenance, and Troubleshooting of Capillary Gas Chromatographic Systems,* second edition, Huthig, Heidelberg, 1995.

8. Scott, R. P. W., *Techniques and Practice of Chromatography,* Dekker, 1995.

Some Representative Non-print Media (See also Chapter 4)

1. *The Sadtler Capillary GC Standard Retention Index Library,* Sadtler, Philadelphia. Computer disk database, contains data on 2000 compounds on four columns.

2. *Pro ezGC,* All, Inc., Dayton, OH; available from lab supply houses. Software for optimizing separations by capillary GC; contains retention index database for 3000 compounds. See review: *Anal. Chem.* **67,** 476A (1995).

3. *On-Line GC Lab,* OnLine Analytics, Inc., Duxbury, MA. Software for modeling and predicting capillary column performance; available with retention index database on 8 columns; See Villalobos, R., and Annino, R., *J. High Resolut. Chromatogr.,* **12,** 149 (1989).

4. *GC-SOS,* The 4S Company, Athens, GA. Simulation software using actual GC data input by the user and then producing simulated chromatograms under different conditions; see review: *Anal. Chem.,* **67,** 360A (1995).

5. *GC Softbook* and *GC Applications and Methods Development Softbook,* The Royal Society, Cambridge, UK; computerized books.

6. *EZChrom,* Scientific Software, Inc., San Ramon, CA. Data handling software.

INDEX OF APPLICATIONS

INDEX